T0140968

Distributed $\mathcal{H}_\infty$ State Estimation with Applications to Multi-Agent Coordination

Von der Fakultät Konstruktions-, Produktions-, und Fahrzeugtechnik
und vom Stuttgart Research Centre for Simulation Technology
der Universität Stuttgart zur Erlangung der Würde eines
Doktor–Ingenieurs (Dr.–Ing.) genehmigte Abhandlung

Vorlegt von

Jingbo Wu
aus Hefei, China

Hauptberichter: Prof. Dr.-Ing. Frank Allgöwer
Mitberichter: Prof. Luca Schenato, Ph.D.
Prof. Dimos Dimarogonas, Ph.D.

Tag der mündlichen Prüfung: 21. April 2017

Institut für Systemtheorie und Regelungstechnik

Universität Stuttgart

2018

Bibliografische Information der Deutschen Nationalbibliothek

Die Deutsche Nationalbibliothek verzeichnet diese Publikation in der Deutschen Nationalbibliografie; detaillierte bibliografische Daten sind im Internet über http://dnb.d-nb.de abrufbar.

D 93

ISBN 978-3-8325-4679-3

Logos Verlag Berlin GmbH
Comeniushof, Gubener Str. 47,
10243 Berlin
Tel.: +49 (0)30 42 85 10 90
Fax: +49 (0)30 42 85 10 92
INTERNET: http://www.logos-verlag.de

Acknowledgment

This thesis contains the results from several years of research work at the Institute for Systems Theory and Automatic Control (IST), University of Stuttgart, Germany. There is a great number of advisors, fellow Ph.D. students, and other people who have contributed to this work or have supported me throughout my time at the institute and beyond:

Most of all, I would like to express my gratitude to my Ph.D advisor, Prof. Frank Allgöwer, for the opportunity to work at the IST, which is a great environment to pursue ones research interests, meet inspiring researchers from all over the world, and have many fruitful discussions. Moreover, it was a place to learn a holistic skill-set including teaching, presenting, and managing projects, which prepares an IST-graduate for professional challenges beyond academic research. Also, I would like to thank Prof. Dimos Dimarogonas and Prof. Luca Schenato for examining my thesis and giving me valuable advice, and Prof. Maike Tilebein for acting as chairwoman of my doctoral committee.

Further, I would like to thank Prof. Valery Ugrinovskii, for hosting me as a guest researcher at the *University of New South Wales* and for several years of fruitful collaboration, while being an important support and advisor for my research work. Also, I would like to thank Prof. Hyungbo Shim and Prof. Brad Yu for hosting me as a guest researcher at the *Seoul National University*, and the *Australian National University* and giving valuable advice.

Also, my colleagues and fellow Ph.D students have been a great support by creating a friendly, collaborative atmosphere at the institute and making every joint project fun and motivating. Most of all, I would like to thank Georg Seyboth, Jan-Maximilian Montenbruck, and Steffen Linsenmayer, for reading the draft of my thesis.

Lastly, special thanks go to my family and my wife Jennifer for their unwavering love and support, which has led me through all difficulties and finally to successfully finishing this work.

Contents

Notation

System variables	
$\mathcal{N}$	The index set $\{1, ..., N\}$ for an integer $N > 0$
$x,\ x_k$	System/agent state
$w,\ w_k$	Disturbance input
$u,\ u_k$	Control input
ϵ_k	Regulation error
z_k	Performance output for synchronization problems
Observer variables	
y_k	Measurement output of observer k
η_k	Measurement noise of y_k
$\hat{x}_k$	Estimate obtained by observer k
e_k	Estimation error of observer k
$\hat{x}_{k,j}$	Estimate of agent state x_j as a component of $\hat{x}_k$
Matrix and norm notation	
I_n	$n \times n$ identity matrix
$A \otimes B$	Kronecker product of two matrices A and B
$A \succ 0$	Matrix A is positive definite
$A \prec 0$	Matrix A is negative definite
$A \succeq 0$	Matrix A is positive semi-definite
$A \preceq 0$	Matrix A is negative semi-definite
$\sigma(A)$	Spectrum of the matrix A
$\lambda_{max}(A)$	Eigenvalue of the matrix A with the largest real part
$\|x\|$	Euclidean norm of the vector x
$\|x\|_W$	Weighted norm $\|x\|_W = \sqrt{x^\top W x}$, with $W \succeq 0$
diag$[A_j]_{j \in I}$	Matrix block-diagonal with $A_{j_1}, ..., A_{j_{\|I\|}}$ on the diagonal for $j \in \mathcal{I}$
$[x_j]_{j \in \mathcal{I}}$	Column vector $[x_{j_1}^\top, ..., x_{j_{\|I\|}}^\top]^\top$ for $\{j_1, ..., j_{j_{\|I\|}}\} = \mathcal{I}$
tr(A)	Trace of the matrix A

Graph notation

$\mathcal{A}$	Adjacency matrix
$\mathcal{L}$	Graph Laplacian matrix
$\mathcal{N}_k$	(Incoming) Neighbourhood of agent k
$\mathcal{M}_k$	(Outgoing) neighbourhood of agent k
p_k	Indegree of agent k
q_k	Outdegree of agent k

Acronyms

ARE	Algebraic Riccati Equation
iSCC	Independent strongly-connected component
LMI	Linear matrix inequality
LTI	Linear time-invariant
MAS	Multi-agent system
RDE	Riccati Differential Equation

Summary

In feedback control theory, observer design is an essential part of many controller design algorithms because in many applications, measured information alone does not sufficiently represent the system's state. Observer design aims at implementing a virtual representation of the system to be controlled, and thus creating a viable estimate of the system's state.

While in the classical observer design one observer is used for one dynamical system, observer schemes have gained attention lately, where multiple observers cooperatively create an estimate of the system's state. These observer schemes are therefore referred to as *distributed estimation*. The essential benefit of distributed estimation lies in the fact that through cooperation, each individual observer only needs very limited sensor capacity, which allows for large, spatially distributed sensor networks. This has the potential to greatly improve estimation performance when the sensors are subject to disturbances, and to enable distributed output feedback control in applications such as large-scale chemical plants, electric grids, and water distribution networks. In such applications, applying a classical observer requires all sensor measurements to be sent to a central fusion unit. In contrast, distributed estimation schemes allow sparse communication topologies, where only neighbouring observer nodes communicate with each other.

However, most distributed estimation schemes in literature suffer from a number of drawbacks, which limit their applicability. For simplicity, many results assume perfect communication channels, the existence of a central coordination unit during the observer design phase, and are restricted to estimating linear systems. Hence, there are many open problems to be solved.

This thesis is dedicated at improving distributed estimation in a number of ways:

- Extending the system class and in particular including nonlinearities,
- Reducing communication bandwidth requirements and energy consumption by implementing event-triggered communication,
- Enabling decentralized computation of the observer parameters by local, iterative optimizations during the observer design phase,
- Preserving scalability of the estimation scheme for systems of increasing size.

Moreover, we apply such cooperating observers to solving the synchronization and output regulation problem for multi-agent systems under difficult circumstances, such as relative measurements, which demonstrates the effectiveness of distributed estimation. For the observers, we achieve $\mathcal{H}_\infty$-type performance guarantees of the estimates with respect to exogeneous disturbances and measurement disturbances. And ultimately, for the closed-loop system, we achieve guaranteed $\mathcal{H}_\infty$-performance of the synchronization or regulation error with respect to these disturbances.

Deutsche Kurzfassung

Verteilte $\mathcal{H}_\infty$ Zustandsschätzung mit Anwendungen auf Multiagenten Systeme

In der Regelungstechnik nimmt die Theorie zu Beobachterentwürfen eine wichtige Rolle ein, da in vielen Anwendungsbeispielen die von der Sensorik gemessene Information allein zum Erreichen der Regelziele nicht ausreicht. Deshalb wird beim Beobachterentwurf ein Modell des zu regelnden Systems implementiert, das geeignete Schätzgrößen für die internen Zustände des Systems liefern soll.

Während beim klassischen Beobachterentwurf ein einzelner Beobachter für ein zu schätzendes System ausgelegt wird, betrachten jüngere Resultat in der Literatur vermehrt den Fall, wo eine Gruppe von Beobachtern gemeinsam den Zustand eines Systems schätzen. Folglich werden solche Verfahren *verteilte Zustandsschätzung* genannt. Der wesentliche Vorteil bei der verteilten Zustandsschätzung liegt darin, dass durch die Kooperation zwischen den einzelnen Beobachtern jeder einzelne Beobachter nur sehr wenig Messinformationen benötigt. Dies ermöglicht die Realisierung großflächiger, räumlich verteilter Sensornetzwerke. Solche Sensornetzwerke können bei Vorliegen von Störungen deren Einfluss bedeutend verringern, und können dafür eingesetzt werden, Ausgangsregler auf verteilter Art und Weise zu implementieren, d.h. Aktuatoren müssen nicht zentral angesteuert werden, sondern können lokal, von getrennten Reglern angesteuert werden. Dies ist besonders relevant bei großen Systemen mit einer Vielzahl von Sensoren und Aktuatoren, wie Chemieanlagen, Stromnetzen und Wasserversorgungsnetzwerken. Klassische Beobachter müssten in diesen Fällen sämtliche Messdaten an einem Ort sammeln und fusionieren, was in räumlich verteilten Systemen hohe Ansprüche an die Kommunikationsstruktur stellt. Verteilte Zustandsschätzung hingegen ermöglicht es, eine dünne Kommunikationsstruktur einzusetzen, wobei nur benachbarte Knoten miteinander kommunizieren.

Die meisten bekannten Resultate zu verteilter Zustandsschätzung nehmen allerdings einige Vereinfachungen an, die die Anwendbarkeit dieser Algorithmen deutlich einschränken. Beispielsweise wird häufig angenommen, dass die Kommunikation störungsfrei und kontinuierlich stattfindet und dass beim Entwurf der verteilten Beobachtern eine zentrale Koordinierung möglich ist, die das gesamte System kennt. Weiterhin wurden nichtlineare Systeme für den Entwurf von verteilten Beobachtern bislang kaum in Betracht gezogen.

Dies vorliegende Arbeit widmet sich der Aufgabe, verteile Zustandsschätzung auf mehrere Art und Weisen weiterzuentwickeln:

- Die Klasse an Systemen, für die verteilte Zustandsschätzung möglich ist, soll erweitert werden. Insbesondere sollen Nichtlinearitäten berücksichtigt werden.

- Ereignisbasierte Kommunikation soll eingeführt werden um die Häufigkeit der Kommunikation deutlich zu verringern. Dies kann sowohl die Anforderungen an die Bandbreite des Kommunikationsnetzwerkes verringern, als auch den einzelnen Beobachtern ermöglichen Energie zu sparen, was vor allem bei batteriebetriebenen Sensoren sinnvoll sein kann.

- Der Entwurf der Beobachter soll ebenfalls auf verteilter Art und Weise ermöglicht werden, damit die Kommunikationsstrukturen wäh-rend der Entwurfsphase und während des Betriebs konsistent sind.

- Skalierbarkeit soll ermöglicht werden in dem Sinne, dass wenn das zu schätzende System in der Größe anwächst, die Komplexität der Beobachter dennoch beschränkt werden kann.

Ferner werden in der vorliegenden Arbeit verteilte Beobachter zur Lösung des Synchronisierungs- und des verteilten Output Regulation-Problems für Multiagenten Systeme eingesetzt. Hier wird der Nutzen dieser Beobachter gezeigt indem relative Messgrößen wie Abstände genutzt werden können um die benötigten Systemzustände zu schätzen. Trotz diesen ungünstigen Messgrößen kann eine $\mathcal{H}_\infty$-Schätzgüte bezüglich Eingangsstörungen und Messstörungen gewährleistet werden, und schlussendlich kann auch eine $\mathcal{H}_\infty$-Regelgüte des geschlossenen Regelkreises erreicht werden.

Chapter 1

Introduction

1.1 Motivation

Recent years have been a time of rapid technological advancement, which is in the process of affecting more and more aspects of our world. Terms like *Digitalization, Internet of Things*, and *Cyber-physical systems* frequently appear in discussions on economy, science, and everyday life (Baheti & Gill, 2011; Manyika et al., 2015). For instance, manufacturing industry is undergoing a tremendous change through new technologies, which are able to sustainably enhance production. In Germany, this transformation is summarized under the term *Industrie 4.0* (Drath & Horch, 2014; Kuka AG, 2016), and some of its key concepts are highly customizable products, efficient supply chain management, and human-robot interaction. As another example, transportation is also experiencing a huge leap, as technologies enabling autonomous driving and platooning are in the process of implementation and can eventually greatly enhance the comfort, efficiency, and safety of private and heavy-duty transportation (Swaroop & Hedrick, 1999; Markoff, 2010; Geiger et al., 2012). Moreover, advances in the coordination of unmanned aerial vehicles (UAVs) can have a huge impact on how existing problems like package delivery, surveillance, and search and rescue missions are tackled (Doherty & Rudol, 2007; Mathew et al., 2015).

All these developments have in common that the systems to be controlled can no longer be considered as isolated individuals, but need to be considered as units interacting with their environment and each other. The realisation of such paradigms is enabled and enhanced by the growth of computational power, availability of communication capabilities, and the increasing amount of sensor data. In particular, communication networks both grow in size and in density, allowing engineers and scientists to tackle increasingly complex and large-scale problems. The contribution of control theory towards these technological advances is the development of rigorous analysis and design methods, which can ensure that large-scale systems exhibit the desired behaviour. Using control theory, suitable interconnection structures can be designed and performance based on the available resources can be optimized. As a result, control theory is experiencing a significant shift of paradigm from the classical single-system/single-controller scheme towards large-scale systems and distributed control (Aström & Kumar, 2014). The essential property of such distributed control schemes is the implementation of a group of individual control units, which may be able to communicate, but make their decisions autonomously. Concerning the exact layout of a group of individual controllers, fundamental questions arise, which need to be addressed:

i) How much information on the large-scale system does each individual controller have?

ii) What is the communication structure and how do the communication channels behave?

iii) Is there any kind of central coordination unit?

Moreover, as control tasks become more and more complex, directly measured information may not be sufficient to meet the control goals. In the classical single-system/single-controller case, observer design methods have been developed to estimate the internal, non-measured states of a complex system, and thus enable *output feedback control*. Two of the most notable early references in this topic are (Kalman, 1960; Luenberger, 1966). Translating these observer design approaches towards large-scale systems and *distributed state estimation* schemes with a group of *distributed observers*, while taking into account the questions above, is a field that is still in its infancy and needs to be further explored.

The results presented in this thesis are dedicated at contributing towards the development of a complete framework for distributed estimation that is able to handle a great variety of system requirements and objectives. The goal is to significantly extend the class of systems that can be addressed and furthermore enhance the practical applicability. In particular, we extend the theory of distributed estimation with respect to multiple directions of complexity, including non-linearities, imperfect communication channels, computational applicability, and applications to multi-agent coordination.

Next, a review on existing results in the field of distributed and cooperative control is given and open questions are pointed out that are addressed in this thesis. Specifically, the literature review will be divided into two parts, in order to distinguish between autonomy of controllers/observers and autonomy of (sub-)systems. In both parts, we will motivate our approach using the technique of *distributed estimation* as part of an output-feedback distributed control setup.

1.2 Distributed control and estimation

The field of distributed control has been an active field of research for several decades, where some important foundations were laid by (Corfmat & Morse, 1976; Šiljak & Vukcevic, 1976; Šiljak, 1978)[1] Many results in this field where collected in (Šiljak, 1991). More recent results were published in (Šiljak et al., 2002; D'Andrea & Dullerud, 2003; Langbort et al., 2004; Stanković et al., 2007; Swigart, 2010), to name a few.

While in the classical controller design one feedback controller is designed to solve the control task, in distributed control, a group of controllers is designed of whom each receives some measurement and determines the input to some actuators. This shift of paradigm is illustrated in Figure 1.1. As the controllers may not receive sufficient measurement information, or the interaction with the other controllers needs to be taken into account, often-times some exchange of information between the controllers is required and thus it is assumed that the individual controllers may send and receive information through some communication network. The significance of the distributed control scheme is that it is able to handle large-scale systems with spatially distributed sensors and actuators much better than a centralized scheme as no central processing entity is needed that collects all measurements centrally.

Distributed controller design however is a complex problem and obtaining suitable control laws always requires the introduction of certain assumptions and eventually leads to a certain amount of conservatism. Some restrictions which are frequently introduced are:

i) It is assumed that the large-scale system can be partitioned, i.e. the system state can be divided into subsystem states and the respective pairs of measurements and actuators belong to one subsystem each.

ii) It is assumed that the communication topology corresponds to the physical interconnections between the subsystems.

[1]The terms *distributed control* and *decentralized control* are not clearly separated in the literature. Nowadays, *decentralized control* is mainly, but not exclusively, used for referring to paradigms where individual controllers do not communicate.

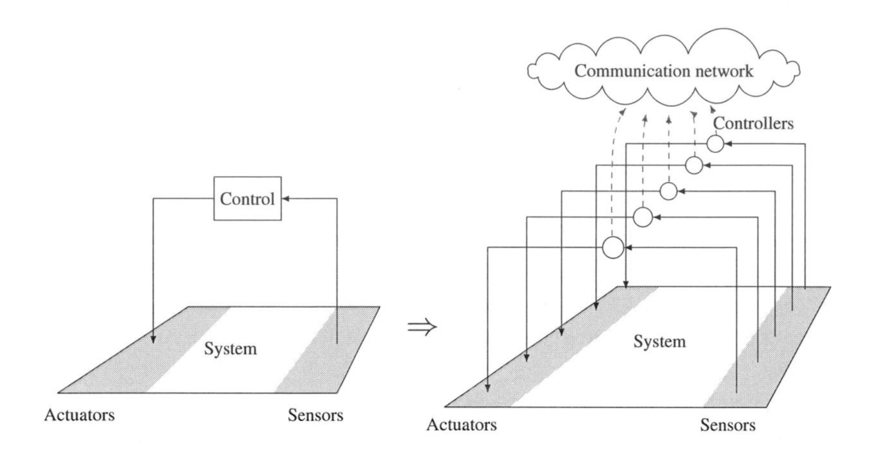

Figure 1.1: Shift of paradigm from single-system-single-controller to distributed control

iii) The class of interconnection topology is restricted to certain types. For instance, cycle-free or string-type interconnections are popular choices.

iv) The controller design is done centrally, i.e. the design conditions may involve all controllers and subsystems.

While these assumptions are often times well justified and viable for the respective control objectives, they still pose some restrictions. Here, the theory of *distributed state estimation* (or short *distributed estimation*) has shown a way to relax the restrictions while preserving the structure of distributed control.

As the name *distributed estimation*[2] suggests, only the estimation part of output feedback control is considered here. While in centralized estimation, all measurements are

[2]Here, it is important to note that with *distributed estimation*, we mean the estimation of the state of a dynamical system. Apart from distributed state estimation, there is a large community, which is concerned with distributed estimation of static parameters, which is not our concern in this thesis. For the same reason, we do not distinguish between *observer* and *estimator*.

fused together in order to determine the state of a dynamical system, in distributed estimation it is proposed that multiple observers each receive some local measurement of lower order. In particular, this reduced measurement may not be sufficient to estimate the system's state. Subsequently, the observers need to cooperate through some communication network, in order to create a viable estimate of the system's state. This becomes particularly challenging in the case of sparse, spatially distributed communication networks. Moreover, even when every single observer may be able to obtain an estimate of the state on its own, through cooperation, the effects of model and measurement disturbances can be significantly reduced (Subbotin & Smith, 2009).

If the distributed estimation problem is solved and thus every individual observer maintains an estimate of the full system state, classical state feedback controller design methods can be applied, while the distributed setup as in Figure 1.1 is still preserved. Therefore, this approach significantly relaxes the restrictions of distributed control described above, while having the drawback of increased computational load for the individual observers/controllers.

Preliminary steps towards this approach have been laid in (Borkar & Varaiya, 1982; Chung & Speyer, 1995), and it has received much attention since the Distributed Kalman Filter was presented in (Olfati-Saber, 2005, 2006, 2007; Carli et al., 2008). More recent work on this topic has been published in (Kamal et al., 2013; Açikmese et al., 2014; de Souza et al., 2016), for instance. For the continuous time case, $\mathcal{H}_\infty$-type distributed estimation was presented in (Ugrinovskii & Langbort, 2011; Ugrinovskii, 2011).

While the literature review above shows that distributed estimation for linear systems has evolved quite far, extensions towards nonlinear systems have rarely been considered. This is in strong contrast to the observer design theory for centralized systems, where numerous approaches have been developed to tackle different classes of nonlinear systems. Observer design for nonlinear systems is especially relevant in the case where the control task does not solely involve the local stabilization of a set point, but rather the convergence with respect to a whole region of attraction, or non-set-point stabilization tasks such as tracking a reference or stabilizing a limit cycle. Therefore, enabling distributed estimation of nonlinear systems is an important goal. Moreover, continuous communication is often assumed in the literature even though it is a significant simplification of the practical circumstances. Hence, in order to make distributed estimation schemes more practicable, communication effects need to be taken into account. In particular, in order to limit the required communication bandwidth and reducing energy-intensive transmission at the same time, event-triggered schemes are highly promising.

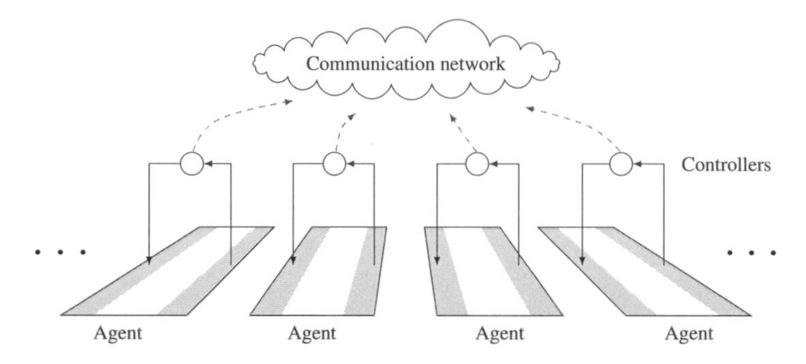

Figure 1.2: Distributed control in the case of autonomous agents

Another question that has to be further investigated is item iv) in the list above. When a set of individual observers are implemented, it is also important to enable the design of those observers in a distributed fashion. Particularly for the case when the whole setup may vary during operation, it is important that the observers are able to adapt to the new circumstances by themselves, which is why effective computational methods need to be developed for distributed observers. Similar goals have been pursued in (Farina & Carli, 2016) for Kalman filtering of dynamically interconnected systems, where it is referred to as *plug-and-play features*. In this thesis, we aim at enabling such kind of plug-and-play operation for distributed estimators as depicted in Figure 1.1.

1.3 Control of autonomous agents

Considering item i) in the previous section, a natural question that arises is which classes of systems incorporate some kind of partitioning in natural way. In fact, the research on so called *Multi-agent systems* (MAS) has received much attention in the last decade, where groups of physically decoupled subsystems are considered, called autonomous *agents*. For this particular case, the distributed control scheme depicted in Figure 1.1 has the very intuitive meaning that every agent has its own local controller, which receives some local measurement and actuates the agent (Figure 1.2).

The "core" of this field of research is the *consensus problem*, which is sometimes also referred to as *agreement problem*. This research direction was initially motivated from particle swarms and animal behaviour (Reynolds, 1987; Vicsek et al., 1995), and has moved far beyond these examples due to its applications to formation control and motion coordination (Jadbabaie & Morse, 2003; Fax & Murray, 2004; Ren et al., 2005, 2007; Ji & Egerstedt, 2007), distributed computation (Xiao et al., 2007; Bolognani et al., 2010), and opinion dynamics (Blondel et al., 2009), to name a few. Other important contributions and surveys were presented in (Moreau, 2005; Olfati-Saber & Murray, 2004; Olfati-Saber et al., 2007; Mesbahi & Egerstedt, 2010). In many of these papers, the presentation of the results is based on graph theory, as the interconnection topology of the agents can be represented by graphs and convergence conditions are subsequently described by using measures for graph connectivity. Moreover, in the classical consensus problem, agents are modelled as integrator systems

$$\dot{x}_k = u_k, \quad k = 1, ..., N$$

and all of the above mentioned references deal with the analysis of the convergence behaviour of such systems under certain input u_k.

Beyond the classical consensus problem with integrator agents, much research was conducted in order to extend the results towards high-order complex agents. Here, the consensus problem is often referred to as *synchronization* as the agents continue to show dynamical behaviour even after reaching agreement[3]. In contrast, in the classical consensus problem, integrator agents converge to a stationary point when reaching consensus. In the case of complex agents one essential question is whether the individual dynamics of the agents are identical or not. In this respect, the MAS is called *homogeneous* or *heterogeneous*. Synchronization of agents modeled as identical linear systems has been thouroughly studied in many papers, including (Fax & Murray, 2004; Tuna, 2008; Scardovi & Sepulchre, 2009), to name a few. Moreover, the literature has been enriched in order to cover various scenarios like time-delayed communication ((Seuret et al., 2008)), quantized measurements ((Frasca et al., 2009)), disturbances ((Ding, 2015)), etc.

In the more general case of heterogeneous MAS, the *Internal Model Principle for Synchronization* was derived in (Wieland & Allgöwer, 2009), which delivers necessary geometric conditions for synchronization. As a matter of fact, every agent in the network has to contain an *Internal Model* that is common to every agent of the multi-agent

[3]in the literature, the word *consensus* is sometimes also used for high-order complex agents. In this thesis, we will refer to *consensus* only for integrator agents.

system. A similar result is presented in (Lunze, 2011), where a leader-follower structure is considered. Moreover, in (Wieland, Sepulchre & Allgöwer, 2011), a constructive method for synchronizing a heterogeneous multi-agent system is presented given that the Internal Model Principle is satisfied. Another approach to tackle heterogeneous MAS is given in (H. Kim et al., 2011) and (X. Wang et al., 2010), where agents are considered as uncertain systems. Since then, further research on output synchronization of heterogeneous MAS has become a very active field with many publications addressing issues such as parameter-varying dynamics (Seyboth et al., 2012), nonlinearites (Isidori et al., 2014; Seyboth & Allgöwer, 2014), and uncertainties (Trentelman et al., 2013).

The results mentioned above established a close connection between synchronization of MAS and the classical *Output Regulation* problem, which is a very general problem description that can incorporate both tracking tasks and disturbance rejection (Francis & Wonham, 1975, 1976; Isidori & Byrnes, 1990; Knobloch et al., 1993). In fact, distributed output regulation is a field that developed in parallel to the literature on synchronization of MAS. Significant references on this topic include (Su & Huang, 2012a, 2012b; Yu & Wang, 2013; Xu et al., 2014; Seyboth et al., 2016). There, the focus does not lie on a group of agents reaching agreement, but instead on the task where a group of agents cooperatively tracks a so called *exosystem* and simultaneously rejects possible disturbances. However, it is fair to say that leader-follower tracking problems without disturbances are the special case where the synchronization problem discussed above and distributed output regulation intersect.

An important aspect when designing synchronizing controllers or regulators is distinguishing whether measurements deliver absolute information or relative information. This concerns both the case of homogeneous and heterogeneous MAS. Here, absolute information means certain measurements $y_k = h_k(x_k)$, which only depend on the local state x_k, where relative information is composed of differences between a number of agents, e.g. $y_{k,j} = h_{k,j}(x_j - x_k)$. For instance, in the task of synchronizing a group of autonomous vehicles, any kind of distance measurement between two agents delivers relative information. Thus, synchronization algorithms need to be developed, which solely rely on this kind of measurements. Some fundamental properties of agents with relative measurements are reviewed in (Barooah & Hespanha, 2007; Zelazo & Mesbahi, 2008).

Moreover, in practical applications, unknown exogeneous disturbances like external forces may affect the agents and measurement uncertainties may diminish the measurement accuracy. Thus, it is of great importance to develop synchronizing controllers, which are able to explicitly take disturbances into account and improve the synchronization performance in this respect.

1.4 Contribution and Outline

Chapter 2: ***Distributed $\mathcal{H}_\infty$ Estimation for Linear and Nonlinear Systems***

The discussion on distributed control and estimation in Section 1.2 shows that there is a considerable number of publications which address distributed estimation for linear systems. However, nonlinear systems have barely been considered. When looking at existing centralized nonlinear estimation algorithms in the literature, one notices that many of them require some kind of transformation a priori. For instance, the Extended Luenberger observer (Zeitz, 1987) and the High-gain Observer (Khalil & Praly, 2014) require a transformation to observability normal form. On the other hand, without coordinate transformation, there are observer design methods in the literature that deal with systems described by a linear state space model with additive nonlinearities (Arcak & Kokotović, 2001b; Açikmese & Corless, 2011).

In this chapter, we elaborate the design of distributed observers for linear systems and discuss various solution methods based on linear matrix inequalities (LMIs). Then, we present an extension towards systems with nonlinearities satisfying incremental quadratic constraints. Our main contributions in this chapter are:

- *We develop a method, which directly extends the LMI-conditions for designing distributed $\mathcal{H}_\infty$ observer for linear systems towards including these nonlinearities.*
- *We discuss the conservativeness of the direct approach by specifically considering the case of monotonous, non-slope-restricted nonlinearities. As it turns out, the resulting matrix equations imposed by this approach are very restrictive in the case where the individual observers receive very limited measurement information.*
- *We develop a further extension of the design conditions, which aims at relaxing this conservativeness. In particular, we propose a two-step algorithm for solving the resulting design conditions.*

Parts of this chapter are based on (Wu & Allgöwer, 2016a).

Chapter 3: ***Event-triggered Communication and Distributed Design of Distributed $\mathcal{H}_\infty$ Observers***

Distributed observers follow some communication structure as depicted in Figure 1.1 and in many results in the literature, no central coordination unit is needed during operation. The task of designing the observers, on the other hand, often needs to be done a priori, and during this design phase, it is often assumed that a central unit is available that gathers all information and pass back the required observer parameters. Furthermore, while the distributed communication structure is retained during operation, it is often assumed for simplicity that the controllers or observers communicate continuously. This is unfavourable in applications due to two reasons: Firstly, in dense communication networks, continuous communication occupies vast bandwidth. And secondly, transmitting information is demanding with respect to energy consumption, which for instance is a drawback in the case of distributed observers in sensor networks that run on a battery.

In this chapter, we aim at enhancing practicability of distributed $\mathcal{H}_\infty$ observers such as those presented in Chapter 2 or (Ugrinovskii, 2011; Wu et al., 2014), by addressing the aforementioned computational and communication issues. Our main contributions in this chapter are:

- *We introduce event-triggered communication between the individual observers, which allows the observers to evaluate whether transmitting information is meaningful or not. This allows the individual observers to greatly reduce the frequency of transmissions, hence saving a large portion of the required communication bandwidth and energy. Here, it is of great importance to find a lower bound for the time intervals between the trigger instances in order to avoid aggregation of transmission events.*

- *We develop distributed design methods for the observers by using distributed optimization technique, including separation of the variables and formulation of a dual design problem. Solving the design conditions can thus be done iteratively and in distributed fashion.*

- *We propose coupled* Riccati Differential Equations *(RDEs) for designing the observer parameters during operation, as an alternative to a priori design of the observer parameters. The basic idea is to randomly initialize the observers and let the required parameters converge online to some suitable values. Here, it is important to show that the RDEs converge to a viable solution although the measurements that an individual observer receives are insufficient for estimating the system's state.*

Parts of this chapter are based on (Wu, Li et al., 2015) and (Wu, Elser et al., 2016).

Chapter 4: ***Distributed $\mathcal{H}_\infty$ State Estimation-based Multi-Agent Coordination***

The starting point of this chapter is the problem of synchronizing heterogeneous MAS based on relative measurements, which we motivated in Section 1.3. The synchronization problem is one of the most considered tasks within the field of multi-agent coordination and using relative measurements instead of absolute measurements means that common sensors like distance sensors can be used for applying synchronizing controllers. The conditions for synchronizing heterogeneous MAS posed by the Internal Model Principle are reviewed and two approaches to designing synchronizing controllers are presented that are discussed with respect to their advantages and limitations. Further, we extend our approach to solving the problem of distributed output regulation, which is another important task within the field of multi-agent coordination. Our main contributions in this chapter are:

- *We develop a geometric approach to solve the synchronization problem, which is based on extending the non-stable modes of the agents to a common Internal Model.*
- *We adapt the* $\mathcal{H}_\infty$ *estimation scheme from Chapter 2 to multi-agent coordination. In particular, scalability needs to be preserved, i.e. the computational complexity of every controller/observer is not supposed to grow to the same extent of the number of agents. Instead, computational demand shall be distributed over a number of agents.*
- *We develop closed-loop synchronization methods that are based on the adapted distributed estimation scheme. In contrast to the geometric approach, the estimation-based approach allows assigning a synchronization trajectory and is able to guarantee* $\mathcal{H}_\infty$*-type synchronization performance with respect to disturbances.*
- *We generalize the problem statement by considering coupled measurements*

$$y_k = h_k(x_k, x_{j_1}, ..., x_{j_p}),$$

which include relative measurements as a special case and furthermore show extensions to synchronizing physically interconnected subsystems. Lastly, we introduce an approach to achieve distributed output regulation based on coupled measurements.

The first part of this chapter including the geometric approach is based on (Wu & Allgöwer, 2012), the results using distributed estimation are based on (Wu et al., 2014; Wu, Ugrinovskii & Allgöwer, 2015, 2016; Wu et al., 2017), and the results for distributed output regulation are based on (Wu & Allgöwer, 2016b).

Chapter 2

Distributed $\mathcal{H}_\infty$ State Estimation for Linear and Nonlinear Systems

This chapter is dedicated to the setup and design of distributed observers for the estimation of systems without making any assumptions on the structure of the system. The characteristic of this estimation problem is that every individual observer maintains an estimate of the system's full state, but in order to achieve a viable estimate, cooperation between the observers is required. In particular, the discussion on distributed control and estimation in the introduction has revealed that there is a considerable number of publications, which address this problem for linear systems. Nonlinear systems, on the other hand, have barely been considered and our goal is to contribute to closing this gap.

In this chapter, we first discuss an $\mathcal{H}_\infty$ approach for designing distributed observers for linear systems similar to (Ugrinovskii, 2011). This approach results in LMI-conditions, which ensure that the distributed observers are well-behaved. Then, we show that the observer scheme and the LMI-conditions can be extended in order to deal with additive nonlinearities satisfying incremental quadratic constraints. What makes this problem particularly challenging is the fact that the vector-storage function approach presented in (Haddad et al., 2004) and used in (Ugrinovskii, 2011) can be shown to be very restrictive as its structural constraints collide with the nonlinearity of the system in general. Subsequently, we pursue an extended observer design method in order to relax these structural constraints.

This chapter is structured as follows: In Section 2.1, the scheme for distributed observers for linear systems is presented and different design algorithms are discussed, including their advantages and drawbacks. Then, in Section 2.2, we discuss observer design methods for nonlinear systems in the centralized case and choose one, which is suitable for extension to the distributed case. Then, the design method for distributed observers is presented in detail, including a thorough discussion on why the classical vector-storage-function approach needs to be extended. Section 2.2 is partially based on (Wu & Allgöwer, 2016a).

2.1 Linear Systems

We consider the LTI system

$$\dot{x} = Ax + B^w w \tag{2.1}$$

where $x(t) \in \mathbb{R}^n$ is the state of the system and w is a $\mathcal{L}_2$-integrable exogenous disturbance with $w(t) \in \mathbb{R}^v$.

As discussed in the introduction, let there be a set of N observers, where each one receives some measurement $y_k(t) \in \mathbb{R}^{r_k}$ from the system (2.1) in the form of

$$y_k = C_k x + \eta_k, \quad k \in \mathcal{N} = \{1, ..., N\}, \tag{2.2}$$

and each of the measurements is subject to a $\mathcal{L}_2$-integrable measurement disturbance η_k.

The vector of stacked measurements y is defined as

$$y = Cx + \eta = \begin{bmatrix} y_1 \\ \vdots \\ y_N \end{bmatrix} = \begin{bmatrix} C_1 x + \eta_1 \\ \vdots \\ C_N x + \eta_N \end{bmatrix}$$

and as a basic requirement for solving the observer design problem, we impose the following assumption.

Assumption 2.1 *The pair (A, C) is detectable.*

However, most importantly, it is not necessary to assume that any individual tuple (A, C_k) is detectable, as lack of detectability of the individual tuples will be compensated for by suitable cooperation among the observers. In particular, it is assumed that the observers are able to communicate through some communication network where the communication topology is represented by a directed graph $\mathcal{G} = (\mathcal{V}, \mathcal{E})$. An overview on graph theoretic basics is given in the Appendix. It should however be noted that we denote edges of a graph in the direction of information flow, i.e. observer k is able to receive data from observer j, if $(v_j, v_k) \in \mathcal{E}$. Further, $\mathcal{N}_k = \{j_1, ..., j_{p_k}\}$ defines the incoming neighbourhood of k, i.e. $(v_j, v_k) \in \mathcal{E}$ for all $j \in \mathcal{N}_k$.

We denote as $[x_j]_{j\in\mathcal{I}}$ the column vector $[x_{j_1}^\top, ..., x_{j_{|I|}}^\top]^\top$, where $\mathcal{I} \subset \mathcal{N}$ is some index set and $\{j_1, ..., j_{j_{|I|}}\} = \mathcal{I}$. With the system dynamics (2.1), the measurement outputs (2.2), and the communication graph $\mathcal{G}$, we arrive at the following problem statement:

Problem 1 (Distributed observers for linear unstructured systems) *For all $k \in \mathcal{N}$, determine a dynamic observer*

$$\dot{\hat{x}}_k = \hat{f}\left(\hat{x}_k, y_k, [\hat{x}_j]_{j\in\mathcal{N}_k}\right), \tag{2.3}$$

which receives its neighbours' estimates $[\hat{x}_j]_{j\in\mathcal{N}_k}$, such that

1. *In the nominal case, i.e. $w \equiv 0, \eta_k \equiv 0$ for all $k \in \mathcal{N}$, we have convergence of the estimates in the sense that*

$$\lim_{t\to\infty} (\hat{x}_k(t) - x(t)) = 0, \tag{2.4}$$

for all initial conditions $\hat{x}_k(0), x(0)$.

2. *For a given positive semi-definite weighting matrix W_k we have $\mathcal{H}_\infty$-type performance of the estimation errors $e_k = x - \hat{x}_k$ in the sense that*

$$\sum_{k=1}^{N} \int_0^\infty e_k^\top W_k e_k dt \leq \gamma^2 \sum_{k=1}^{N} \int_0^\infty (\|w\|^2 + \|\eta_k\|^2) dt + I_0, \tag{2.5}$$

where I_0 is the cost due to the observers' uncertainty about the initial condition of the system (2.1).

The first part of Problem 1 represents the classical observer design problem in analogy to (Luenberger, 1966). In particular, if the pair (A, C_k) is detectable for all $k \in \mathcal{N}$, then this part is solvable without communication between the observers. The second part of Problem 1 represents an $\mathcal{H}_\infty$-type performance inequality, which characterizes the observers' estimation error with respect to both the exogeneous disturbance w and the measurement disturbances η_k. In particular, in the sense of $\mathcal{H}_\infty$-control, the performance parameter γ should be minimized in order to improve the observers' performance.

2.1.1 Observer scheme

An elemental form for distributed observers is given as

$$\dot{\hat{x}}_k = A\hat{x}_k + L_k(y_k - C_k\hat{x}_k) + K_k \sum_{j\in\mathcal{N}_k} (\hat{x}_j - \hat{x}_k), \tag{2.6}$$

with estimates $\hat{x}_k(t) \in \mathbb{R}^n$ and matrices of suitable dimension L_k, K_k. This form of distributed observer has been presented in (Ugrinovskii, 2011) and was recently further generalized in (L. Wang & Morse, 2017). Many of our results throughout this thesis will be based on variations of this observer scheme. For the sake of simplicity, we assume that the communication channels are not affected by disturbances. However, disturbances such as additive noise or time delays can be taken into account, which will be discussed later.

With the definition of the observers (2.6), Problem 1 becomes the design of the matrices L_k and K_k for all $k \in \mathcal{N}$, such that (2.4) and (2.5) are satisfied. In fact, the observer dynamics (2.6) is composed of a classical Luenberger observer (Luenberger, 1966) and a linear consensus term relying on the deviation between the local estimates of an observer and the estimates of its neighbours. Therefore, we will refer to L_k as the *correction gain matrix* and K_k as the *consensus gain matrix* and together as the *observer gain matrices*.

The dynamics of the observer errors e_k can be calculated with (2.1) and (2.6), which yields

$$\dot{e}_k = (A - L_k C_k) e_k - L_k \eta_k + B^w w + K_k \sum_{j \in \mathcal{N}_k} (e_j - e_k). \tag{2.7}$$

If every pair (A, C_k) is detectable, then Problem 1 can be solved individually by setting $K_k = 0$ and designing L_k using standard methods from centralized observer design. But since (A, C_k) is not assumed to be detectable in general, cooperation between the observers needs to be taken into account.

Remark 2.1 *Note that instead of using the observer gain matrix K_k in front of the summed diffusive term $\sum_{j \in \mathcal{N}_k} (\hat{x}_j - \hat{x}_k)$, the coupling between the observers can be modified as $\sum_{j \in \mathcal{N}_k} K_{kj} (\hat{x}_j - \hat{x}_k)$. This provides the observer scheme with additional degrees of freedom, which can be used in order to improve performance. In later sections, we will use the latter scheme, but for now we keep the simplified version for the sake of clarity of notation.*

Digression: Intuition for the case with a leader node

An interesting special case arises if the communication graph $\mathcal{G}$ is connected and contains a leader node v_1, i.e. v_1 does not receive any data by communication. Then, v_1 is the root of a spanning tree such as depicted in Figure 2.1 and has no incoming edges. In order to solve Problem 1, the pair (A, C_1) clearly needs to be detectable. Moreover, in this case, we can derive the following results on the nominal case of Problem 1.

Lemma 2.1 *Let the pair (A, C_1) be detectable and let v_1 be the leader node of the connected communication graph $\mathcal{G}$. Then the convergence property 2.4 is achieved in the nominal case with $w \equiv 0, \eta_k \equiv 0$ for all $k \in \mathcal{N}$, if*

- *$A - L_1C_1$ is Hurwitz and*
- *$K_k = \kappa I_n, L_k = 0$ for $k = 2, ..., N$, where $A - Re(\lambda_2)\kappa I$ is Hurwitz and λ_2 is the eigenvalue of the Laplacian matrix $\mathcal{L}$ with the smallest nonzero real part.*

Proof. *The stacked error system with $e = [e_j]_{j\in\mathcal{N}}$ can be written as*

$$\dot{e} = \begin{bmatrix} A - L_1C_1 & 0 \\ 0 & I_{N-1} \otimes A \end{bmatrix} e - (\mathcal{L} \otimes \kappa I_n)\, e.$$

Since v_1 is a leader node, we have

$$\mathcal{L} = \begin{bmatrix} 0 & 0 \\ * & \overline{\mathcal{L}} \end{bmatrix}$$

with the submatrix $\overline{\mathcal{L}} \in \mathbb{R}^{(N-1)\times(N-1)}$. Thus, the stacked system can be rewritten as

$$\dot{e} = \begin{bmatrix} A - L_1C_1 & 0 \\ * & I_{N-1} \otimes A - \overline{\mathcal{L}} \otimes \kappa I_n \end{bmatrix} e.$$

Now, we choose the transformation $\Delta = T^{-1}\overline{\mathcal{L}}T$ with an upper-triangular matrix Δ. Since $\mathcal{G}$ is connected, all eigenvalues of $\overline{\mathcal{L}}$, i.e. all diagonal elements of Δ, have positive real part (cf. (Mesbahi & Egerstedt, 2010)). With

$$e = \left(\begin{bmatrix} 1 & 0 \\ 0 & T \end{bmatrix} \otimes I_n \right) v,$$

the error system can be transformed to

$$\begin{aligned} \dot{v} &= \begin{bmatrix} A - L_1C_1 & 0 \\ * & I_{N-1} \otimes A - \Delta \otimes \kappa I_n \end{bmatrix} v \\ &= \left[\begin{array}{c|ccc} A - L_1C_1 & 0 & \dots & 0 \\ \hline * & A - \lambda_2\kappa I_n & & * \\ * & & \ddots & \\ * & 0 & & A - \lambda_N\kappa I_n \end{array} \right] v, \end{aligned}$$

where $\lambda_2, ..., \lambda_N$ are the nonzero eigenvalues of $\mathcal{L}$. Since $A - Re(\lambda_2)\kappa I$ is Hurwitz, we have that $A - Re(\lambda_k)\kappa I$ is Hurwitz for all $k = 2, ..., N$. Therefore, we have $v \to 0$ and thus, $e \to 0$. ■

As a matter of fact, as $\mathrm{Re}(\lambda_2) > 0$, one can always find $\kappa > 0$ such that $A - \mathrm{Re}(\lambda_2)\kappa I$ is Hurwitz. This Lemma shows that in the leader-follower case with one observer receiving sufficient measurements, the problem can be solved using graph theoretic methods as in (Fax & Murray, 2004). For the cycle-free case, the result can be further simplified.

Lemma 2.2 *Let the pair (A, C_1) be detectable and let v_1 be the leader node of the connected communication graph $\mathcal{G}$. Moreover, let the graph $\mathcal{G}$ be cycle-free. Then the convergence property 2.4 is achieved in the nominal case with $w \equiv 0, \eta_k \equiv 0$ for all $k \in \mathcal{N}$, if*

- *$A - L_1 C_1$ is Hurwitz and*
- *$K_k = \kappa I_n$ for $k = 2, ..., N$, where $A - \kappa I$ is Hurwitz.*

Proof. *This lemma follows from Lemma 2.1 and the observation that $\overline{\mathcal{L}}$ is a lower-triangular matrix with a suitable ordering of the vertices.* ■

The Lemmas 2.1 and 2.2 show a simple design method under the assumption that there exists an observer, which receives sufficient measurement information, and without taking performance into account. The measurements y_k of the other observers are simply discarded. Moreover, if there is a central coordination unit, then Lemma 2.2 can always be applied instead of Lemma 2.1, since a connected graph $\mathcal{G}$ can always be reduced into a cycle-free connected graph, as depicted in Figure 2.1.

Moving back to the general case with no leader node and Problem 1 including the performance guarantees, more effort needs to be spend in order to find suitable gain matrices L_k and K_k. We consider the storage-function component

$$V_k(e_k) = e_k^\top P_k e_k, \tag{2.12}$$

where $P_k \in \mathbb{R}^{n \times n}$ is a symmetric, positive definite matrix. Thus, the Lie derivative of the component $V_k(e_k)$ is

$$\dot{V}_k(e_k) = 2e_k^\top P_k(A_k - L_k C_k)e_k + 2e_k^\top P_k(-L_k\eta_k + B^w w) + 2e_k^\top P_k K_k \sum_{j \in \mathcal{N}_k} (e_j - e_k). \tag{2.13}$$

Let p_k be the in-degree of the vertex v_k with respect to the communication graph $\mathcal{G}$ and $\alpha_k > 0$ an arbitrarily small constant. Then, with the substitutions

$$\begin{aligned} P_k L_k &= G_k \\ P_k K_k &= F_k \end{aligned} \tag{2.14}$$

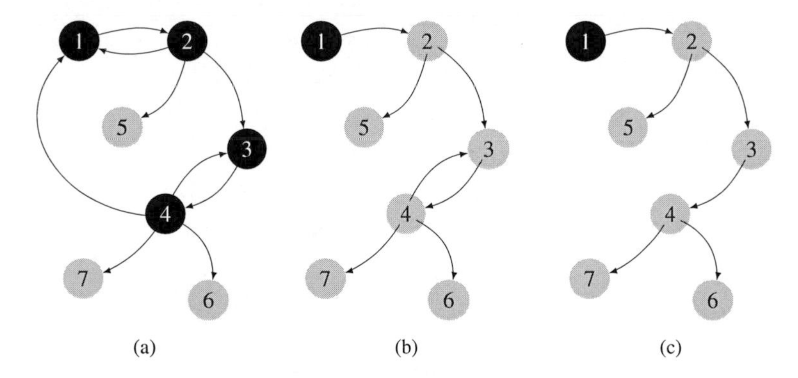

Figure 2.1: Connected communication graph $\mathcal{G}$. Roots are filled black. By cancellation of two edges, node 1 is turned into a leader (Subfigure (b)). By cancellation of another edge, the graph is made cyle-free (Subfigure (c)).

and the matrix definition

$$Q_k = P_k A + A^\top P_k - G_k C_k - C_k^\top G_k^\top - p_k F_k - p_k F_k^\top + \alpha_k P_k, \tag{2.15}$$

the Lie derivative (2.13) is simplified to

$$\dot{V}_k = e_k^\top Q_k e_k - \alpha_k e_k^\top P_k e_k + 2e_k^\top P_k(-L_k \eta_k + B^w w) + 2e_k^\top P_k K_k \sum_{j \in \mathcal{N}_k} e_j \tag{2.16}$$

In order to satisfy property 2 of Problem 1, we aim at obtaining a dissipation inequality of the form

$$\sum_{k=1}^{N} \dot{V}_k \leq \gamma^2 \sum_{k=1}^{N} (\|w\|^2 + \|\eta_k\|^2) - \sum_{k \in \mathcal{N}} e_k^\top W_k e_k,$$

where the derivation of the storage-function components (2.16) will be used in order to find suitable conditions for designing the observers. Next, we describe two variants for finding such conditions and discuss their structural differences.

2.1.2 Centralized LMI-based solution

As a centralized approach to solving Problem 1, one can derive the design condition in the form of an LMI by stacking the estimation errors $e = [e_j]_{j \in \mathcal{N}}$ and using the storage function candidate $V = e^\top \text{diag}[P_j]_{j\in\mathcal{N}} e$. The Lie derivative of V then equals the sum of the component-wise derivatives (2.16), i.e.

$$\begin{aligned}
\dot{V} =& \sum_{k=1}^{N} \dot{V}_k \\
=& \sum_{k=1}^{N} e_k^\top Q_k e_k - \sum_{k=1}^{N} \alpha_k e_k^\top P_k e_k + \sum_{k=1}^{N} 2e_k^\top P_k(-L_k\eta_k + B^w w) + \sum_{k=1}^{N} 2e_k^\top P_k K_k \sum_{j\in\mathcal{N}_k} e_j \\
\prec& \sum_{k=1}^{N} e_k^\top Q_k e_k - \sum_{k=1}^{N} 2e_k^\top G_k\eta_k + \sum_{k=1}^{N} 2e_k^\top P_k B^w w + \sum_{k=1}^{N} 2e_k^\top F_k \sum_{j\in\mathcal{N}_k} e_j \\
=& e^\top diag[Q_j]_{j\in\mathcal{N}} e + e^\top diag[F_j]_{j\in\mathcal{N}}((\mathcal{A}^\top + \mathcal{A}) \otimes I_n)e - \sum_{k=1}^{N} 2e_k^\top G_k\eta_k + \sum_{k=1}^{N} 2e_k^\top P_k B^w w
\end{aligned}$$

The design LMI for satisfying Properties 1 and 2 or Problem 1 is shown in the following Theorem.

Theorem 2.1 (Centralized conditions for linear systems) *Let a set of matrix variables* $G_k, F_k, P_k, k \in \mathcal{N}$ *and* $\gamma > 0$ *be a solution to*

$$\left[\begin{array}{c|c|c} diag[Q_j + W_j]_{j\in\mathcal{N}} + diag[F_j]_{j\in\mathcal{N}}((\mathcal{A}^\top + \mathcal{A}) \otimes I_n) & -diag[G_j]_{j\in\mathcal{N}} & [P_j B^w]_{j\in\mathcal{N}} \\ \hline * & -\gamma^2 I_r & 0 \\ \hline * & 0 & -\gamma^2 I_v \end{array}\right] \preceq 0, \tag{2.17}$$

where $r = \sum_{k=1}^{N} r_k$ *and* $\mathcal{A}$ *is the adjacency matrix of the communication graph* $\mathcal{G}$. *Then, the distributed observers* (2.6) *with the gain matrices* (2.14) *are a solution to Problem 1.*

This LMI is centralized in the sense that it requires knowledge about the entire system including all output matrices C_k and its size given by $(Nn + r + v) \times (Nn + r + v)$ scales with the number of observers. In practice, this requires some sort of bidirectional star-shaped communication structure where the central unit computes the observer gains for all observers and feeds them back to the entire system.

The storage function candidate $V = e^\top \text{diag}[P_j]_{j\in\mathcal{N}} e$ introduces a certain amount of conservatism in general by using the block-diagonal matrix $\text{diag}[P_j]_{j\in\mathcal{N}}$ instead of a full-block matrix P. For certain classes of systems, conservatism can be excluded, such as for block-triangular systems in the nominal case for $w, \eta_k = 0$ (Carlson et al., 1992). However, note that block-triangularity in the case of the state vector $e = [e^{1\top}, ..., e^{N\top}]^\top$ with (2.7) essentially means that the communication graph $\mathcal{G}$ needs to be cycle-free, i.e. the case discussed in Lemma 2.2. Due to this conservatism, LMI-conditions such as (2.17) are sufficient conditions, but, in general, not necessary for solvability of Problem 1.

Remark 2.2 *The parameters α_k determine the prescribed convergence speed. Therefore, large α_k enforces faster convergence, but leads to large observer gains and thus diminishes performance with respect to measurement noise.*

2.1.3 Distributed LMI-based solution

Considering the centralized solution in the previous section, we notice that gathering all data of the entire system is not aligned with the basic idea of distributed control and estimation of having a set of cooperative observers that interact with their neighbours only.

In order to enable observer design in a distributed fashion, it is meaningful to propose the design conditions as a set of N LMI conditions. Specifically, defining these LMI conditions should only require local knowledge and the dimension of the LMIs should only become larger when more neighbours are added. In order to achieve this, we extend (2.16) by adding and subtracting slack variables $\widetilde{P}_k$, such that

$$\begin{aligned}\dot{V}_k(e) =&e_k^\top Q_k e_k - \alpha_k e_k^\top P_k e_k + 2e_k^\top P_k(-L_k\eta_k + B^w w) + 2e_k^\top P_k K_k \sum_{j\in\mathcal{N}_k} e_j \\ &+ q_k e_k^\top \widetilde{P}_k e_k - q_k e_k^\top \widetilde{P}_k e_k,\end{aligned}$$

where q_k is the out-degree of the vertex v_k with respect to the communication graph $\mathcal{G}$. Summing up the slack terms, it holds that

$$\sum_{k=1}^{N} q_k e_k^\top P_k e_k = \sum_{k=1}^{N} \sum_{j\in\mathcal{N}_k} e_j^\top P_j e_j \tag{2.18}$$

and subsequently, if the dissipation inequality

$$\dot{V}_k(e) \leq \gamma^2(\|w\|^2 + \|\eta_k\|^2) - e_k^\top(W_k + \alpha_k P_k)e_k - q_k e_k^\top \widetilde{P}_k e_k + \sum_{j\in\mathcal{N}_k} e^{j\top} \widetilde{P}_j e_j \tag{2.19}$$

is satisfied for all $k \in \mathcal{N}$, then it holds for the sum $V = \sum_{k\in\mathcal{N}} V_k$ that

$$\dot{V} = \sum_{k=1}^{N} \dot{V}_k \leq \gamma^2 \sum_{k=1}^{N}(\|w\|^2 + \|\eta_k\|^2) - \sum_{k\in\mathcal{N}} e_k^\top(W_k + \alpha_k P_k)e_k. \tag{2.20}$$

With these derivations, we conclude the following Theorem, which is a variation of the result from (Ugrinovskii, 2011), (Ugrinovskii & Langbort, 2011).

Theorem 2.2 (Distributed conditions for linear systems) *Let a collection of matrices* F_k, $\widetilde{P}_k \succ 0$, G_k *and* $P_k \succ 0$, $k \in \mathcal{N}$, *be a solution of the LMIs*

$$\left[\begin{array}{ccc:c} Q_k + W_k + q_k\widetilde{P}_k & -G_k & P_kB^w & \mathbf{1}_{p_k}^\top \otimes F_k \\ -G_k^\top & -\gamma^2 I & 0 & 0 \\ B^{w\top}P^{k\top} & 0 & -\gamma^2 I & 0 \\ \hdashline \mathbf{1}_{p_k} \otimes F_k^\top & 0 & 0 & -diag[\widetilde{P}_j]_{j\in\mathcal{N}_k} \end{array}\right] \preceq 0 \tag{2.21}$$

for all $k \in \mathcal{N}$, *then the observers* (2.6) *with the observer gain matrices*

$$L_k = P_k^{-1}G_k$$
$$K_k = P_k^{-1}F_k.$$

solve Problem 1.

Proof. *The proof for this Theorem follows from the fact that satisfaction of the LMI* (2.21) *leads to the dissipation inequality* (2.20). *The nominal case with* $w, \eta_k = 0$ *implies exponential stability of the origin since*

$$\dot{V} \leq -\sum_{k\in\mathcal{N}} \alpha_k V_k,$$

i.e. Property 1 of Problem 1. And integrating (2.20) *on both sides yields the inequality*

$$\int_0^T \dot{V}dt \leq \gamma^2 \sum_{k=1}^N \int_0^T (\|w\|^2 + \|\eta_k\|^2)dt - \sum_{k\in\mathcal{N}} \int_0^T e_k^\top(W_k + \alpha_k P_k)e_k dt$$
$$\sum_{k\in\mathcal{N}} \int_0^T e_k^\top W_k e_k dt \leq \gamma^2 \sum_{k=1}^N \int_0^T (\|w\|^2 + \|\eta_k\|^2)dt + V(0).$$

This corresponds to Property 2 of Problem 1 with $I_0 = V(0)$. *Thus, Problem 1 is solved.*
■

Following Theorem 2.2, the existence of a solution of the LMIs (2.21) is sufficient for solvability of Problem 1. The resulting communication topology between the estimators is represented by the graph $\mathcal{G}$. In particular, all coupling terms are diffusive, i.e. differences of the estimates $\hat{x}_k$. Hence, in the nominal case of $w, \eta_k = 0$, the estimates $\hat{x}_k$ converge to the system state x and all coupling terms between the observers converge to 0.

It should be noted that with larger number of observers, more LMIs need to be solved. However, for all $k \in \mathcal{N}$, the respective LMI only grows in its dimension if more neighbours are added to the observer, or when the system dimension n grows.

Remark 2.3 *The LMIs are coupled by the slack variables $\widetilde{P}_k, k \in \mathcal{N}$, which prevent the LMIs to be solved in a purely decentralized fashion. However, the structure of the LMIs* (2.21) *allows for separation of the problem in the sense that solving the LMIs is amenable to distributed optimization algorithms. This will be addressed in Chapter 3.*

Remark 2.4 *The parameter γ in* (2.21) *can be set as a fixed value or can be minimized by substituting $\gamma^2 = \beta > 0$ and minimize β subject to the LMI constraints* (2.21). *Since* (2.21) *define a convex set, this is a tractable problem.*

2.1.4 Relation to vector dissipativity

The slack variables $\widetilde{P}_k$ in the previous section do not need to be handled as separate variables. Instead, specific values can be assigned a priori or the dimension of the solution can be reduced. One interesting case arises if we set the slack variables as $\widetilde{P}_k = \pi_k P_k$ with some scalar $\pi_k > 0$, as this has a close relation to vector dissipativity theory.

Let $\widetilde{P}_k = \pi_k P_k$ with some scalar $\pi_k > 0$, then the component dissipation inequality (2.19) can be interpreted in terms of vector dissipativity. For this purpose, we consider the vector storage function

$$V_{\text{vec}} = \begin{bmatrix} V_1 \\ \vdots \\ V_N \end{bmatrix}.$$

Vector storage functions are thoroughly discussed in (Haddad et al., 2004). In our case, for analysing the Lie-derivative of V_{vec}, we can apply (2.19) and thus obtain

$$\dot{V}_{\text{vec}} = \begin{bmatrix} \dot{V}_1 \\ \vdots \\ \dot{V}_N \end{bmatrix} = \begin{bmatrix} \gamma^2(\|w\|^2 + \|\eta_1\|^2) - e_1^\top W_1 e_1 \\ \vdots \\ \gamma^2(\|w\|^2 + \|\eta_N\|^2) - e_N^\top W_N e_N \end{bmatrix} + \Pi V_{\text{vec}}, \tag{2.22}$$

with

$$\Pi = \mathcal{A} \begin{bmatrix} \pi_1 & & \\ & \ddots & \\ & & \pi_N \end{bmatrix} - \begin{bmatrix} q_1\pi_1 + \alpha_1 & & \\ & \ddots & \\ & & q_N\pi_N + \alpha_N \end{bmatrix} \tag{2.23}$$

Further, we show the following property for Π.

Lemma 2.3 *The matrix Π defined in* (2.23) *is Hurwitz for all $\pi_1, ..., \pi_N > 0$.*

Proof. *Since the out-degree q_k is defined as the number of edges $(v_k, v_j) \in \mathcal{E}$ for $j \in \mathcal{N}$, we know that for all $k \in \mathcal{N}$*

$$q_k + \frac{\alpha_k}{\pi_k} > \sum_{j=1}^{N} \mathcal{A}_{jk}$$

and therefore

$$q_k \pi_k + \alpha_k > \sum_{j=1}^{N} \pi_k \mathcal{A}_{jk},$$

which means that Π is diagonal-dominant with negative entries on the diagonal. From the theory of Gershgorin disks (Feingold et al., 1962), we thus conclude that Π is Hurwitz. ■

Due to Lemma 2.3, vector dissipativity holds with storage function V_{vec} and vector supply rate

$$S(e_1, ..., e_N, w, \eta_1, ..., \eta_N) = \begin{bmatrix} \gamma^2(\|w\|^2 + \|\eta_1\|^2) - e^{1\top} W_1 e_1 \\ \vdots \\ \gamma^2(\|w\|^2 + \|\eta_N\|^2) - e^{N\top} W_N e_N \end{bmatrix}.$$

Then, for the nominal case with $w, \eta_k = 0$, exponential convergence of $V_{\text{vec}} \to 0$ holds. Moreover, by integrating (2.22) on both sides, Property 2 of Problem 1 can be shown. Thus, Problem 1 is solved.

Using this vector dissipativity approach with $\widetilde{P}_k = \pi_k P_k$ to achieve a solution of Problem 1 requires the parameters π_k to be set a priori. Due to increased generality, using the slack-variables $\widetilde{P}_k$ can yield better performance values γ. However, $\widetilde{P}_k = \pi_k P_k$ reduces the number of variables, which has computational advantages and will be assumed in most results throughout the remaining thesis.

2.2 Nonlinear Systems

Considering existing nonlinear estimation algorithms in the literature, one notices that many of them require some kind of transformation a priori. For instance, the *Extended Luenberger Observer* (Zeitz, 1987) and the *High-gain Observer* (Gauthier et al., 1992; Khalil & Praly, 2014) require a transformation to observability normal form. However, in the case when multiple observers cooperate in a distributed setup and each of them has its individual sensor capabilities, then transforming the coordinates diminishes the observers' ability to create the estimate in a cooperative manner. Requiring the state transformation to be the same for all observers however requires the sensor capabilities to be essentially the same, which is highly restrictive. Thus, transformation-based observer design method have very limited applicability in a distributed setup. On the other hand, without coordinate transformation, there are observer design methods in the literature that deal with systems described by a linear state space model with additive nonlinearities, which satisfy certain conditions such as globally Lipschitz conditions (Thau, 1973; Raghavan & Hedrick, 1994), sector-bounds (Arcak & Kokotović, 2001b), and incremental quadratic constraints (Açikmese & Corless, 2011). The latter in fact generalizes the first two cases, which is why we will focus on this class of nonlinearities.

We consider the nonlinear system

$$\dot{x} = Ax + B^{\phi}\phi(Hx) + B^{w}w, \tag{2.24}$$

where $x(t) \in \mathbb{R}^n$ is the state of the system and w is an $\mathcal{L}_2$-integrable exogenous disturbance with $w(t) \in \mathbb{R}^v$.

As in the linear case, a group of N observers each measure an output $y_k(t) \in \mathbb{R}^{r_k}$ of the form

$$y_k = C_k x + D_k \phi(Hx), k \in \mathcal{N} \tag{2.25}$$

In the linear case (2.1) we did not include the control input for the sake of simplicity. In the nonlinear case (2.24), however, the separation principle does not hold in general. Therefore, we need to make the following technical assumption.

Assumption 2.2 *Given the initial condition $x(0)$ then $x(t)$ exists and is unique for all $t \geq 0$.*

The same assumption was made in (Arcak & Kokotović, 2001b). Note that we did not include measurement noise into y_k since in the general case of non-slope-restricted nonlinearities $\phi(\cdot)$, we cannot provide performance guarantees such as (2.5). For the special case of slope-restricted nonlinearities, the extension with measurement noise can easily be established.

In both (2.24) and (2.25), $\phi : \mathbb{R}^{l_1} \to \mathbb{R}^{l_2}$ is a known nonlinearity satisfying the incremental quadratic constraint

$$\begin{bmatrix} z_2 - z_1 \\ \phi(z_2) - \phi(z_1) \end{bmatrix}^{\top} M \begin{bmatrix} z_2 - z_1 \\ \phi(z_2) - \phi(z_1) \end{bmatrix} \geq 0 \tag{2.26}$$

for all $z_1, z_2 \in \mathbb{R}^{l_1}$, with some symmetric matrix $M \in \mathbb{R}^{(l_1+l_2)\times(l_1+l_2)}$. In the remainder of this section, we consider the class $M \in \mathcal{M}$ defined as

$$\mathcal{M} = \left\{ \begin{bmatrix} X & Z \\ Z^\top & -Y \end{bmatrix} \middle| X = X^\top \succeq 0, Y = Y^\top \succeq 0 \right\}. \tag{2.27}$$

Moreover, in many practical applications, $x(t)$ will be restricted to a bounded set $\mathcal{X}$. In this case, it suffices for (2.26) to hold on $\mathcal{X}$.

Now, the distributed estimation problem can be posed as follows:

Problem 2 (Distributed observers for nonlinear unstructured systems) *For all $k \in \mathcal{N}$, determine a dynamic observer*

$$\dot{\hat{x}}_k = \hat{f}\left(\hat{x}_k, y_k, [\hat{x}_j]_{j\in\mathcal{N}_k}\right), \tag{2.28}$$

which receives its neighbours' estimates, such that

1. *In the nominal case, i.e. $w \equiv 0$, we have convergence of the estimates in the sense that*
$$\lim_{t\to\infty} (\hat{x}_k(t) - x(t)) = 0, \tag{2.29}$$
for all initial conditions $\hat{x}_k(0), x(0)$.

2. *For a given positive semi-definite weighting matrix W_k the observers provide $\mathcal{H}_\infty$-type performance of the estimation errors $e_k = x - \hat{x}_k$ in the sense that*
$$\frac{1}{N}\sum_{k=1}^{N}\int_0^\infty e_k^\top W_k e_k dt \le \gamma^2 \sum_{k=1}^{N}\int_0^\infty \|w\|^2 dt + I_0, \tag{2.30}$$
where I_0 is the cost due to the observers' uncertainty about the initial condition of the system (2.24).

2.2.1 Observer scheme

The observer dynamics are proposed to be

$$\begin{aligned} \dot{\hat{x}}_k =& A\hat{x}_k + B^\phi\hat{\phi}_k + L_k(y_k - \hat{y}_k) + \sum_{j\in\mathcal{N}_k} K_{kj}(\hat{x}_j - \hat{x}_k) \\ \hat{y}_k =& C_k\hat{x}_k + D_k\hat{\phi}_k, \end{aligned} \tag{2.31}$$

where $\hat{\phi}_k$ is a nonlinear function satisfying the equation

$$\hat{\phi}_k = \phi\left(H\hat{x}_k + \widetilde{L}_k(y_k - C_k\hat{x}_k - D_k\hat{\phi}_k) + \sum_{j\in\mathcal{N}_k}\widetilde{K}_{kj}(\hat{x}_j - \hat{x}_k)\right). \tag{2.32}$$

The observer gains to be designed are $L_k, \widetilde{L}_k, K_{kj}$, and $\widetilde{K}_{kj}$, which are real matrices of suitable dimension. Thus, for all $k \in \mathcal{N}$, these matrices need to be determined such that the two properties of Problem 2 are satisfied simultaneously.

Note that in the distributed observers (2.6), we had the correction gain matrices L_k and the consensus gain matrices K_k. The consensus gain matrices are now individualized in order to provide for additional degrees of freedom and moreover, $\widetilde{L}_k$ and $\widetilde{K}_{kj}$ are introduced as additional correction gain matrices in the estimate of the nonlinearity (2.32). As it will later turn out, in the case of slope restricted nonlinearities, these matrices $\widetilde{L}_k$ and $\widetilde{K}_{kj}$ can be set to 0, which significantly simplifies observer design at the expense of performance.

As it can be seen in (2.32), there is an algebraic loop with respect to $\hat{\phi}_k$. This needs to be resolved in order to implement (2.31). Therefore, we impose the following assumption:

Assumption 2.3 *For all $k \in \mathcal{N}$, there exists a continuous function $\hat{\phi}_k(\hat{x}_k, \hat{x}_{j,j\in\mathcal{N}_k})$, which uniquely solves Equation* (2.32).

Clearly, (2.32) is solvable if $D_k = 0$. If $D_k \neq 0$, then $\hat{\phi}_k$ needs to be explicitly calculated.

Remark 2.5 *A similar assumption was made in (Açikmese & Corless, 2011), where the authors go even further and discuss the general case, when there is another algebraic loop in the nonlinearity ϕ itself. This extension can also be investigated in the case of distributed estimation, but will not be discussed further in this thesis.*

Before delving into the design of the observer gain matrices, we look at the incremental quadratic constraint (2.26) and develop inequalities, which follow from it with respect to the estimation errors. First, let $\delta\phi_k = \phi - \hat{\phi}_k$. Then, for the observer error, we obtain with (2.24) and (2.31) that

$$\begin{aligned}\dot{e}_k =& Ae_k + B^\phi\delta\phi_k - L_k(C_ke_k + D_k\delta\phi_k) + \sum_{j\in\mathcal{N}_k} K_{kj}(e_j - e_k) + B^w w\\ =&(A - L_kC_k - \sum_{j\in\mathcal{N}_k} K_{kj})e_k + \sum_{j\in\mathcal{N}_k} K_{kj}e_j + B^w w + (B^\phi - L_kD_k)\delta\phi_k. \end{aligned}\tag{2.33}$$

The incremental-nonlinearity $\delta\phi_k$ can be re-written as

$$\begin{aligned}
\delta\phi_k =& \phi(z_2) - \phi(z_1) \\
z_2 =& Hx \\
z_1 =& H\hat{x}_k + \widetilde{L}_k(y_k - C_k\hat{x}_k - D_k\hat{\phi}_k) + \sum_{j\in\mathcal{N}_k} \widetilde{K}_{kj}(\hat{x}_j - \hat{x}_k).
\end{aligned} \tag{2.34}$$

Here, the benefits of the additional observer gain matrices $\widetilde{L}_k$ and $\widetilde{K}_{kj}$ become apparent, as the incremental-nonlinearity, which affects (2.33), can be shaped as needed. Thus, with (2.26), it holds that

$$\begin{bmatrix} z_2 - z_1 \\ \delta\phi_k \end{bmatrix}^\top M \begin{bmatrix} z_2 - z_1 \\ \delta\phi_k \end{bmatrix} \geq 0$$

where

$$z_2 - z_1 = He_k - \widetilde{L}_k(C_k e_k + D_k\delta\phi_k) - \sum_{j\in\mathcal{N}_k} \widetilde{K}_{kj} e_k + \sum_{j\in\mathcal{N}_k} \widetilde{K}_{kj} e_j.$$

Subsequently, the incremental quadratic constraint can be reformulated as

$$\begin{bmatrix} e_k \\ \delta\phi_k \\ [e_j]_{j\in\mathcal{N}_k} \\ w \end{bmatrix}^\top \Phi_k^\top M \Phi_k \begin{bmatrix} e_k \\ \delta\phi_k \\ [e_j]_{j\in\mathcal{N}_k} \\ w \end{bmatrix} \geq 0 \tag{2.35}$$

with

$$\Phi_k = \left[\begin{array}{cc:ccc:c} H - \widetilde{L}_k C_k - \sum_{j\in\mathcal{N}_k} \widetilde{K}_{kj} & -\widetilde{L}_k D_k & \widetilde{K}_{kj_1} & \dots & \widetilde{K}_{kj_{p_k}} & 0 \\ 0 & I & 0 & \dots & 0 & 0 \end{array}\right]. \tag{2.36}$$

The reformulated incremental quadratic constraints (2.35) now describe a relation between specific, system-related variables, which appear in the estimation error dynamics (2.33). Therefore, the inequalities (2.35) are particularly valuable for constructing suitable LMI-conditions for designing the observer gain matrices using S-procedure.

2.2.2 Observer design by vector storage functions

Our first approach to solve Problem 2 is adapted from the approach presented in Section 2.1, by introducing a vector storage function candidate of the form

$$V_{\text{vec}} = \begin{bmatrix} V_1(e_1) \\ \vdots \\ V_N(e_N) \end{bmatrix}, \tag{2.37}$$

where $V_k(e_k) = e_k^\top P_k e_k$. In the following, we will discuss the capabilities and restrictions of this approach.

In order to solve Problem 2 we construct distributed LMI-conditions that guarantee vector dissipativity in the sense of

$$\dot{V}_{\text{vec}} \leq \Pi V_{\text{vec}} + S(e, w), \tag{2.38}$$

where Π is Hurwitz and $S(e, w)$ is some vector supply rate subject to the estimation error $e = [e_1^\top \; \dots \; e_N^\top]^\top$ and disturbance w. We define the matrices

$$Q_k = P_k A + A^\top P_k - G_k C_k - (G_k C_k)^\top - \sum_{j\in\mathcal{N}_k} F_{kj} - \sum_{j\in\mathcal{N}_k} F_{kj}^\top + \alpha_k P_k + q_k \pi_k P_k,$$

where $P_k \in \mathbb{R}^{n\times n}$ is a symmetric, positive definite matrix. The parameters $\pi_k > 0$ are set in order to define the desired dissipation property (2.38). The parameters $\alpha_k > 0$ may be set arbitrarily small and determines the convergence speed. Then, we have the following Theorem.

Theorem 2.3 *Let a collection of matrices F_k, G_k and P_k, $k \in \mathcal{N}$, be a solution of the LMIs*

$$\left[\begin{array}{cc|ccc|c} Q_k + W_k & P_k B^\phi - G_k D_k & F_{kj_1} & \dots & F_{kp_k} & P_k B^w \\ * & 0 & 0 & 0 & 0 & 0 \\ \hline * & 0 & -\pi_{j_1} P_{j_1} & 0 & 0 & 0 \\ * & 0 & 0 & \ddots & 0 & 0 \\ * & 0 & 0 & 0 & -\pi_{j_{p_k}} P_{j_{p_k}} & 0 \\ \hline * & 0 & 0 & 0 & 0 & -\gamma^2 I \end{array}\right] + \Phi_k^\top M \Phi_k \preceq 0 \tag{2.39}$$

with $\gamma > 0$, where $\mathcal{N}_k = \{j_1, \dots, j_{p_k}\}$ and $W_k \geq 0$, then Problem 1 admits a solution of the form (2.31), *where*

$$\begin{aligned} L_k &= P_k^{-1} G_k \\ K_{kj} &= P_k^{-1} F_{kj}. \end{aligned} \tag{2.40}$$

Proof. *The proof of this Theorem is omitted here as the extended design introduced in section 2.2.4 generalizes this result.* ■

Theorem 2.3 is a preliminary theorem, where it is assumed that $\widetilde{L}_k$ and $\widetilde{K}_{kj}$ are determined a priori. In fact, for certain classes of multipliers M, this is a viable procedure, as the shape of M defines the choice for $\widetilde{L}_k$ and $\widetilde{K}_{kj}$. One example for this is the circle criterion observer (Arcak & Kokotović, 2001b), which in its distributed form will be discussed in the next section. Moreover, a method for simultaneously designing all observer gain matrices including $\widetilde{L}_k$ and $\widetilde{K}_{kj}$ can be found by factorizing $\Phi_k^\top M \Phi_k$, which will be discussed later.

2.2.3 Conservatism of the vector storage function

As an important example, we consider element-wise monotonous nonlinearities $\phi : \mathbb{R}^r \to \mathbb{R}^r$ without slope restriction as in (Arcak & Kokotović, 2001b), given as

$$\phi(Hx) = \begin{bmatrix} \phi_1(\sum_{j=1}^n H_{1j}x_j) \\ \vdots \\ \phi_r(\sum_{j=1}^n H_{rj}x_j) \end{bmatrix},$$

where each $\phi_i, i = 1, ..., r$ satisfies the incremental sector condition

$$(a-b)^\top (\phi_i(a) - \phi_i(b)) \geq 0.$$

This class of nonlinearities ϕ satisfies the incremental quadratic constraint (2.26) with

$$M = \begin{bmatrix} 0 & I \\ I & 0 \end{bmatrix}. \tag{2.41}$$

The difference $\phi(z_2) - \phi(z_1)$ can be expressed in terms of a time-varying nonlinearity $\psi(z_2 - z_1, t)$, where each entry ψ_i satisfies the sector constraint $z\psi_i(z,t) \geq 0, \forall z \in \mathbb{R}$. For $r = 1$, this is depicted in Figure 2.2.

Let us for now consider the case, where the nonlinearity does not affect the output y_k, i.e. $D_k = 0$ for all $k = 1, ..., N$. The following matrix-definitions are made in this section:

$$\begin{aligned} A_k &= A - L_k C_k - \sum_{j \in \mathcal{N}_k} K_{kj} \\ A_{kj} &= K_{kj} \\ E_k &= H - \widetilde{L}_k C_k - \sum_{j \in \mathcal{N}_k} \widetilde{K}_{kj} \\ E_{kj} &= \widetilde{K}_{kj} \quad k \in \mathcal{N} \end{aligned} \tag{2.42}$$

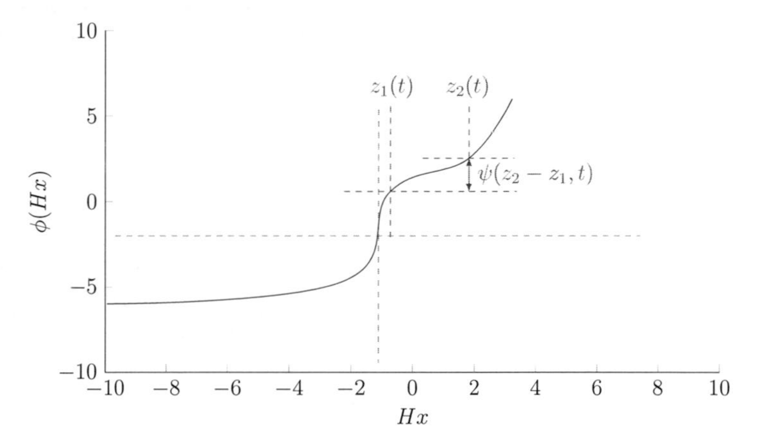

Figure 2.2: Example for monotonous, non-slope-restricted nonlinearities.

Then, the error systems (2.33) with $w = 0$ can be expressed as N interconnected systems

$$\begin{aligned}\dot{e}_k &= A_k e_k + \sum_{j\in\mathcal{N}_k} A_{kj} e_j + B^{\phi} v_k \\ \zeta_k &= E_k e_k + \sum_{j\in\mathcal{N}_k} E_{kj} e_j,\end{aligned} \tag{2.43}$$

with the feedback loops $v_k = \psi_k(\zeta_k, t) \in [0, \infty]$. Thus, we can compose the matrices

$$\widetilde{A} = \begin{bmatrix} A_1 & A_{12} & \dots & A_{1N} \\ A_{21} & \ddots & & \vdots \\ \vdots & & & \\ A_{N1} & \dots & & A_{NN} \end{bmatrix}, \quad E = \begin{bmatrix} E_1 & E_{12} & \dots & E_{1N} \\ E_{21} & \ddots & & \vdots \\ \vdots & & & \\ E_{N1} & \dots & & E_{NN} \end{bmatrix}, \quad B = I_N \otimes B^{\phi}, \tag{2.44}$$

in order to write the interconnected systems (2.43) as a system

$$\begin{aligned}\dot{e} &= \widetilde{A} e + B v \\ \zeta &= E x,\end{aligned} \tag{2.45}$$

where we use the stacked vectors $e = [e_j]_{j\in\mathcal{N}}, v = [v_j]_{j\in\mathcal{N}}, \zeta = [\zeta_j]_{j\in\mathcal{N}}$.

The stability of the closed-loop system 2.45 with the feedback $v = \psi(\zeta, t)$ can be directly related to the theory of absolute stability (see e.g. (Narendra, 2014) and Chapter 7 in (Khalil, 2001)). Some of the early references to this problem are (Popov, 1961; Kalman,

1963; Willems, 1973). Two of the main results for solving the problem of absolute stability are the *Circle Criterion* and the *Popov Criterion*. The main difference between these two approaches lies in the Lyapunov function candidate, which is a quadratic function $V = x^\top Px$ in the case of the Circle Criterion, and a Lur'e-type function of the form $V = \frac{1}{2}x^\top Px + \sum_{k\in\mathcal{N}} \gamma_k \int_0^{\zeta_k} \psi_k(\sigma, t)d\sigma$ in the case of the Popov Criterion. More recent results, which further analysed these methods were published in (Pittet et al., 1997; Suykens et al., 1998; Bliman, 2000; Arcak & Kokotović, 2001a; Arcak et al., 2003).

Here, applying the Circle Criterion approach tells us that the origin of the closed-loop system (2.45) is globally asymptotically stable, if the transfer function matrix $G(s) = -E(sI - \widetilde{A})^{-1}B$ is strictly positive real (SPR). Further, from the Kalman-Yakubovich-Popov (KYP) Lemma (Rantzer, 1996), we have following necessary and sufficient conditions.

Lemma 2.4 *Let* $(\widetilde{A}, B)$ *be controllable and* $(\widetilde{A}, E)$ *be observable. Then, the transfer function matrix* $G(s) = -E(sI - \widetilde{A})^{-1}B$ *is strictly positive real if and only if there exists a symmetric matrix* $P > 0$ *and a constant* $\alpha > 0$ *such that*

$$\begin{aligned} P\widetilde{A} + \widetilde{A}^\top P &\preceq -\alpha I \\ PB &= -E^\top. \end{aligned} \tag{2.46}$$

Note that, since $B = I_N \otimes B^\phi$, (2.46) implies that P can only have block-diagonal form $P = P_{\text{diag}} = \text{diag}(P_1, ..., P_N)$ if E is also block-diagonal, i.e. $E_{kj} = \widetilde{K}_{kj} = 0$ for all k, j.

However, the vector storage function candidate V_{vec} used in Theorem 2.3 corresponds to a block-diagonal solution $P = P_{\text{diag}}$ of (2.46), as in both case, no mixed-terms are included. This relation can be formalized in the following Lemma:

Lemma 2.5 *If there exists collection of matrices* $F_k, G_k, P_k, k \in \mathcal{N}$*, which solves the LMIs in Theorem 2.3 with* M *defined in* (2.41)*, then* $P = \text{diag}(P_1, ..., P_N)$ *solves* (2.46)*.*

Proof. *The proof to this Lemma can be established by considering the quadratic Lyapunov function* $V = \sum_{k\in\mathcal{N}} e_k^\top P_k e_k$ *and applying the LMI* (2.39)*.* ■

We conclude that the solvability of (2.46) is a necessary condition for the existence of a solution of Theorem 2.3. With $E_{kj} = \widetilde{K}_{kj} = 0$ for all k, j, the diagonal entries can be rewritten as

$$P_k B^\phi = -H^\top + C_k^\top \widetilde{L}_k^\top \tag{2.47}$$

for all $k \in \mathcal{N}$. Since in a distributed estimation setup it is generally assumed that C_k has low rank and in particular (A, C_k) is not necessarily observable, (2.47) is highly conservative. This is illustrated with a simulation example later in this section. In the following, we investigate extending the approach presented in Theorem (2.3) towards relaxing the conservatism discussed here.

2.2.4 Extended observer design

The discussions in Section 2.2.3 indicate that using the vector storage function candidate (2.37) introduces a significant amount of conservatism to the solution. In the following, we derive alternative LMI-conditons in order to relax this issue. The conditions presented in this section are build upon the idea of relaxing the block-diagonal structure of P in (2.46). Again, the design conditions are desired to be of distributed form in order to enable the application of distributed optimization algorithms for designing the observers.

Note that in the remainder of this section, we make the assumption of bidirectional graphs. Extensions to directed graphs can be established, but result in more complex LMIs.

First, we introduce the matrix definitions

$$\begin{aligned}
Q_{k,k} &= P_k(A - L_kC_k - \sum_{j\in\mathcal{N}_k} K_{kj}) + \alpha_k P_k + (A - L_kC_k - \sum_{j\in\mathcal{N}_k} K_{kj})^\top P_k + p_k\pi_k P_k + W_k \\
Q_{k,j} &= P_k K_{kj} + (A - L_kC_k - \sum_{i\in\mathcal{N}_k} K_{ki})^{\top} P_{kj} \qquad (2.48) \\
S^k_{j,i} &= P_{kj}^\top K_{ki} + K_{kj}^\top P_{ki} \\
\widetilde{B}_k &= B^\phi - L_k D_k \\
\Phi_k &= \left[\begin{array}{cc|ccc|c} H - \widetilde{L}_kC_k - \sum_{j\in\mathcal{N}_k} \widetilde{K}_{kj} & -\widetilde{L}_kD_k & \widetilde{K}_{kj_1} & \dots & \widetilde{K}_{kj_{p_k}} & 0 \\ 0 & I & 0 & \dots & 0 & 0 \end{array}\right]
\end{aligned}$$

which are needed for the following Theorem on the design of distributed observers.

Theorem 2.4 (Distributed observers for nonlinear systems) *Consider a nonlinear system* (2.24) *and the set of observers* (2.31), *where the communication topology is represented by the undirected graph $\mathcal{G}$. For all $k \in \mathcal{N}$, let the matrices P_k, and P_{kj} with $P_{kj} = P_{jk}^\top$ for $j \in \mathcal{N}_k$ and $P_{kj} = 0$ for $j \notin \mathcal{N}_k$. satisfy*

$$P = \begin{bmatrix} P_1 & P_{12} & \dots & P_{1N} \\ P_{21} & P_2 & & P_{2N} \\ \vdots & & \ddots & \vdots \\ P_{N1} & P_{N2} & \dots & P_N \end{bmatrix} \succ 0. \tag{2.49}$$

Then, let the collection of matrices L_k, $\widetilde{L}_k$, and K_{kj}, $\widetilde{K}_{kj}$ be a solution to the matrix inequalities

$$\begin{aligned} &\left[\begin{array}{cc|cccc|c} Q_{k,k} & P_k\widetilde{B}_k & Q_{k,j_1} & \dots & Q_{k,j_{p_k}} & & P_k B^w \\ * & 0 & \widetilde{B}_k^\top P_{kj_1} & \dots & \widetilde{B}_k^\top P_{kj_{p_k}} & & 0 \\ \hline * & * & S^k_{j_1,j_1} & \dots & S^k_{j_1,j_{p_k}} & & P_{kj_1}^\top B^w \\ * & * & * & \ddots & \vdots & & \vdots \\ * & * & * & * & S^k_{j_{p_k},j_{p_k}} & & P_{kj_{p_k}}^\top B^w \\ \hline * & * & * & * & * & & -\gamma^2 I \end{array}\right] \\ &+ \left[\begin{array}{cc|ccc|c} 0 & 0 & & 0 & & 0 \\ 0 & 0 & & 0 & & 0 \\ \hline & & -\pi_{j_1}P_{j_1} & & 0 & \\ 0 & 0 & & \ddots & & 0 \\ & & 0 & & -\pi_{j_{p_k}}P_{j_{p_k}} & \\ \hline 0 & 0 & & 0 & & 0 \end{array}\right] + \Phi_k^\top M \Phi_k \preceq 0, \end{aligned} \tag{2.50}$$

with $\{j_1, ..., j_{p_k}\} = \mathcal{N}_k$. Then, the observers (2.31) *are a solution to Problem 2 in the sense of* (2.30).

The design conditions presented in this theorem are not immediately applicable as positive definiteness of P is a centralized condition and the matrix inequalities contain the quadratic form $\Phi_k^\top M \Phi_k$. However, this Theorem serves as a basis for developing a numerically efficient method for solving Problem 1.

Proof. We use the storage function candidate

$$V(e) = \sum_k \left(e_k^\top P_k e_k + \sum_{j\in\mathcal{N}_k} e_k^\top P_{kj} e_j \right), \tag{2.51}$$

where non-zero mixed-terms $e_k^\top P_{kj} e_j$ are added in order to relax the structural constraints imposed in Section 2.2.2. For the Lie-derivative of $V(e)$ we have

$$\dot{V}(e) = \sum_{k=1}^N \left(e_k^\top P_k \dot{e}_k + \dot{e}_k^\top P_k e_k\right) + \sum_{k=1}^N \left(\sum_{j\in\mathcal{N}_k} e_k^\top P_{kj}\dot{e}_j + \dot{e}_k^\top P_{kj} e_j \right).$$

From the fact that $\mathcal{G}$ is undirected we observe that for every $(v_k, v_j) \in \mathcal{E}$ we have both $e_k^\top P_{kj}\dot{e}_j$ and $e_j^\top P_{jk}\dot{e}_k$ as parts of $\dot{V}$. In addition, $P_{jk} = P_{kj}^\top$. Therefore, by replacing $e_k^\top P_{kj}\dot{e}_j$ with $e_j^\top P_{kj}^\top \dot{e}_k$, we obtain

$$\dot{V}(e) = \sum_{k=1}^{N} (e_k^\top P_k \dot{e}_k + \dot{e}_k^\top P_k e_k) + \sum_{k=1}^{N} \left(\sum_{j\in\mathcal{N}_k} e_j^\top P_{kj}^\top \dot{e}_k + \dot{e}_k^\top P_{kj} e_j \right)$$

With (2.33), we obtain the terms

$$\begin{aligned}
&e_k^\top P_k \dot{e}_k \\
&= e_k^\top P_k (A - L_k C_k - \sum_{j\in\mathcal{N}_k} K_{kj}) e_k + e_k^\top P_k \left(\sum_{j\in\mathcal{N}_k} K_{kj} e_j + B^w w + (B^\phi - L_k D_k)\delta\phi_k \right), \\
&e_j^\top P_{kj}^\top \dot{e}_k \\
&= e_j^\top P_{kj}^\top (A - L_k C_k - \sum_{i\in\mathcal{N}_k} K_{ki}) e_k + e_j^\top P_{kj}^\top \left(\sum_{i\in\mathcal{N}_k} K_{ki} e_j + B^w w + (B^\phi - L_k D_k)\delta\phi_k \right).
\end{aligned}$$

Subsequently, by the definitions (2.48), the Lie-derivative of $V(e)$ can be written as

$$\begin{aligned}
\dot{V}(e) = &\sum_{k=1}^{N} \left(e_k^\top Q_{k,k} e_k + 2 e_k^\top P_k (B^w w + \widetilde{B}_k \delta\phi_k) - (\alpha_k + p_k\pi_k) e_k^\top P_k e_k - e_k^\top W_k e_k \right) \\
&+ \sum_{k=1}^{N} \sum_{j\in\mathcal{N}_k} (e_k^\top Q_{k,j} e_j + e_j^\top Q_{k,j}^\top e_k) + \sum_{k=1}^{N} \sum_{i,j\in\mathcal{N}_k} e_j^\top S_{j,i}^k e_i \\
&+ \sum_{k=1}^{N} \sum_{j\in\mathcal{N}_k} e_j^\top P_{kj}^\top (B^w w + \widetilde{B}_k \delta\phi_k).
\end{aligned}$$

With (2.50), we now have

$$\begin{aligned}
\dot{V} \leq &\sum_{k=1}^{N} \left(-e_k^\top (p_k \pi_k P_k + \alpha_k P_k + W_k) e_k + \sum_{j\in\mathcal{N}_k} e_j^\top \pi_j P_j e_j \right) + \gamma^2 N w^\top w \\
\leq &- \min_k \alpha_k e^\top e - \sum_{k=1}^{N} e_k^\top W_k e_k + \gamma^2 N w^\top w. \qquad (2.52)
\end{aligned}$$

Since $W \geq 0$ and $\alpha_k > 0$, we have exponential convergence of $e \to 0$ for $w = 0$. Integrating (2.52) over $[0, \infty)$ yields the desired $\mathcal{H}_\infty$-performance

$$\int_0^\infty \dot{V}(e)dt + \sum_{k=1}^{N} \int_0^\infty e_k^\top W_k e_k dt < \int_0^\infty N\gamma^2 w^\top w dt$$
$$\frac{1}{N}\sum_{k=1}^{N} \int_0^\infty e_k^\top W_k e_k dt < \gamma^2 \|w\|_{\mathcal{L}_2}^2 + I_0,$$

for $I_0 = V(e(0))$. ■

Theorem 2.4 provides conditions for distributed observers that guarantee robust performance with respect to input disturbances. However, as mentioned, the matrix inequalities (2.50) are not linear in the solution variables and the condition of $P \succ 0$ is a centralized one. Therefore, we need to find an efficient solution strategy for calculating filter gains that satisfy the conditions of Theorem 2.4.

In the following, we step-by-step transform the design conditions to a computationally tractable problem. Firstly, we introduce distributed LMI-conditions for positive definiteness of P.

Lemma 2.6 *The matrix P defined in* (2.49) *is positive definite, if for all $k \in \mathcal{N}$ it holds that*

$$P^{(k)} = \left[\begin{array}{c|ccc} \frac{1}{1+p_k}P_k & \frac{1}{2}P_{kj_1} & \dots & \frac{1}{2}P_{k,j_{p_k}} \\ \hline * & \frac{1}{1+p_{j_1}}P_{j_1} & & 0 \\ * & & \ddots & \\ * & 0 & & \frac{1}{1+p_{j_{p_k}}}P_{j_{p_k}} \end{array}\right] \succ 0, \tag{2.53}$$

where $\{j_1, ..., j_{p_k} = \mathcal{N}_k\}$.

Proof. *Let $x = [x_1^\top \ \dots \ x_N^\top]^\top$, where $x_k \in \mathbb{R}^n$ for $k \in \mathcal{N}$. Moreover, let P be defined as* (2.49) *with $P_{kj} = 0$ for $j \notin \mathcal{N}_k$. Then, this lemma can be proven by decomposing the quadratic form as*

$$x^\top P x = \sum_{k=1} x_k^\top P_k x_k + \sum_{k=1}^{N} x_k^\top \sum_{j \in \mathcal{N}_k} P_{kj} x_j$$
$$= \sum_{k=1}^{N} \left(\sum_{j \in \mathcal{N}_k \cup k} \frac{1}{1+p_j} x_j^\top P_j x_j + x_k^\top \sum_{j \in \mathcal{N}_k} P_{kj} x_j \right).$$

Due to (2.53)*, it is ensured that the term on the right-hand side is positive for all x, which implies positive definiteness of P.* ■

In order to apply (2.53), the in-degrees of the neighbours, $p_{j_1}...p_{j_{p_k}}$ are needed. Note that this requires only local information, whereas global knowledge about the complete graph is not necessary. However, if the exact degrees are unkown, it also suffices to replace any p_j with an upper bound $p_{max} \geq p_j$.

Next, we denote

$$\Phi_k = \begin{bmatrix} \Phi_k^{(1)} \\ \Phi_k^{(2)} \end{bmatrix} = \left[\begin{array}{cc|ccc|c} H - \widetilde{L}_k C_k - \sum_{j \in \mathcal{N}_k} \widetilde{K}_{kj} & -\widetilde{L}_k D_k & \widetilde{K}_{kj_1} & \dots & \widetilde{K}_{kj_{p_k}} & 0 \\ \hline 0 & I & 0 & \dots & 0 & 0 \end{array}\right] \tag{2.54}$$

and factorize the term $\Phi_k^\top M \Phi_k$ with (2.27) as follows.

Lemma 2.7 *Given any symmetric matrix C. Let $X > 0$, then $C + \Phi_k^\top M \Phi_k$ is negative semi-definite if and only if*

$$\begin{bmatrix} C + \Phi_k^{(1)\top} Z \Phi_k^{(2)} + \Phi_k^{(2)\top} Z^\top \Phi_k^{(1)} - \Phi_k^{(2)\top} Y \Phi_k^{(2)} & \Phi_k^{(1)\top} \\ \Phi_k^{(1)} & -X^{-1} \end{bmatrix} \preceq 0.$$

Otherwise, let $X = 0$, then $C + \Phi_k^\top M \Phi_k$ is negative semi-definite if and only if

$$C + \Phi_k^{(1)\top} Z \Phi_k^{(2)} + \Phi_k^{(2)\top} Z^\top \Phi_k^{(1)} - \Phi_k^{(2)\top} Y \Phi_k^{(2)} \preceq 0.$$

Proof. The quadratic form can be factorized as

$$\begin{aligned} & C + \begin{bmatrix} \Phi_k^{(1)\top} & \Phi_k^{(2)\top} \end{bmatrix} \begin{bmatrix} X & Z \\ Z^\top & -Y \end{bmatrix} \begin{bmatrix} \Phi_k^{(1)} \\ \Phi_k^{(2)} \end{bmatrix} \\ =& C + \begin{bmatrix} \Phi_k^{(1)\top} & \Phi_k^{(2)\top} \end{bmatrix} \begin{bmatrix} X\Phi_k^{(1)} + Z\Phi_k^{(2)} \\ Z^\top \Phi_k^{(1)} - Y\Phi_k^{(2)} \end{bmatrix} \\ =& C + \Phi_k^{(1)\top} X \Phi_k^{(1)} + \Phi_k^{(1)\top} Z \Phi_k^{(2)} + \Phi_k^{(2)\top} Z^\top \Phi_k^{(1)} - \Phi_k^{(2)\top} Y \Phi_k^{(2)} \end{aligned}$$

and then, the lemma follows by applying the Schur-complement on X if $X \succ 0$. ■

Remark 2.6 *Note that Lemma 2.7 did not explicitly deal with positive semi-definite $X \succeq 0$. However, if* (2.26) *is satisfied by $M \in \mathcal{M}$, it is also satisfied by*

$$
\begin{array}{c}
\qquad\qquad\textbf{I} \qquad\qquad\qquad\qquad\qquad\qquad\qquad \textbf{II} \qquad\qquad\qquad\qquad \textbf{III} \qquad \textbf{IV} \\
\left[\begin{array}{cc:ccc:c:c}
\widetilde{Q}_{k,k} & T_k+(H-\widetilde{L}_k C_k-\sum_{j\in\mathcal{N}_k}\widetilde{K}_{kj})^\top Z & \widetilde{Q}_{k,j_1} & \dots & \widetilde{Q}_{k,j_{p_k}} & P_k B^w & \\
* & -Z^\top \widetilde{L}_k D_k - D_k^\top \widetilde{L}_k^\top Z - Y & \widetilde{T}_{k,j_1}+Z^\top \widetilde{K}_{kj_1} & \dots & \widetilde{T}_{k,j_{p_k}}+Z^\top \widetilde{K}_{kj_{p_k}} & 0 & \\
\hdashline
* & * & \widetilde{S}^k_{j_1,j_1}-\pi_{j_1}P_{j_1} & \dots & S^k_{j_1,j_{p_k}} & P_{kj_1}^\top B^w & \Phi_k^{(1)\top} \\
* & * & * & \ddots & \vdots & \vdots & \\
* & * & * & * & \widetilde{S}^k_{j_{p_k},j_{p_k}}-\pi_{j_{p_k}}P_{j_{p_k}} & P_{kj_{p_k}}^\top B^w & \\
\hdashline
* & * & * & * & * & -\gamma^2 I & \\
\hdashline
 & & * & & & & -X^{-1}
\end{array}\right] \preceq 0
\end{array}
\qquad (2.55)
$$

$$
\widetilde{M} = M + \begin{bmatrix} X + \xi I & Z \\ Z & -Y \end{bmatrix},
$$

with $\xi > 0$. Hence, X can be replaced by $X + \xi I$, which makes it suitable for applying Lemma 2.7.

The Lemmas 2.6 and 2.7 provide suitable conditions for expressing the positive definiteness of P and dealing with the quadratic term $\Phi_k^\top M \Phi_k$. Concluding, we can state following result for solving (2.49) and (2.50).

Corollary 2.1 *Let $X > 0$. The matrix inequality* (2.50) *holds if* (2.55) *is satisfied with $\widetilde{Q}_{k,k} = Q_{k,k}$, $\widetilde{Q}_{k,j} = Q_{k,j}$, $\widetilde{S}^k_{j,i} = S^k_{j,i}$, $T_k = P_k \widetilde{B}_k$ and $\widetilde{T}_{k,j} = \widetilde{B}^\top P_{kj}$.*

Proof. *The corollary follows from Lemma 2.7 by applying* (2.54). ■

Remark 2.7 *The matrix inequality* (2.55) *is composed modularly with the following components:*

I Estimation core *problem.*

II Local interactions with neighbors. This part is omitted in the centralized ($N = 1$) case.

III Performance with respect to the disturbance.

IV Factorization of $\Phi_k^{(1)\top} X \Phi_k^{(1)}$. This part is omitted if $X = 0$.

With Corollary 2.1, we can resolve the quadratic terms $\Phi_k^\top M \Phi_k$ and replace the centralized condition $P \succ 0$ with distributed conditions. It is now left to find a solution method that can deal with product terms like $P_k L_k$, $P_{kj} L_k$.

2.2.5 Numerical calculation for the extended observer design

While in Theorem 2.3, the substitutions $P_k L_k = G_k$ and $P_k K_k = F_k$ were made to make the problem convex in the solution variables, in the extended case this does not solve the issue directly due to the off-diagonal blocks $P_{kj}, j \in \mathcal{N}_k$. The problem has been discussed in many papers on decentralized control, e.g. (Scherer, 2002; Stanković et al., 2007; Swigart, 2010). Due to this non-convexity, there is no general solution method available, but instead, alternative design methods are required. While Youla-Parametrization as in (Scherer, 2002; Swigart, 2010) is not suitable for the class of interconnection graphs under consideration, using a robustness argument as in (Stanković et al., 2007) has proven to be too conservative in the present context. In the following we will present a two-step solution strategy that can efficiently solve the matrix inequalities (2.55).
Step 1: Set the desired performance $\gamma > 0$ and solve

$$\begin{aligned} &\min \sum_{(v_k, v_j) \in \mathcal{E}} \|P_{kj}\| \\ &\text{subject to (2.53), (2.55),} \end{aligned} \tag{2.56}$$

where (2.55) is defined with

$$\begin{aligned} \widetilde{Q}_{k,k} &= P_k A - G_k C_k - \sum_{j \in \mathcal{N}_k} F_{kj} + A^\top P_k - C_k^\top G_k^\top - \sum_{j \in \mathcal{N}_k} F_{kj}^\top + \alpha_k P_k + p_k \pi_k P_k + W_k, \\ \widetilde{Q}_{k,j} &= F_{kj} + A^\top P_{kj} - \lambda_k C_k^\top C_k P_{kj} - \lambda_k p_k P_{kj}, \\ \widetilde{S}_{j,i}^k &= \lambda_k P_{kj}^\top + \lambda_k P_{ki}, \\ T_k &= P_k (B^\phi - L_k D_k), \\ \widetilde{T}_{k,j} &= B^{\phi\top} P_{kj} - \lambda_k D_k^\top C_k^\top P_{kj}. \end{aligned} \tag{2.57}$$

Here, G_k and F_{kj} are matrix variables of suitable dimension and $\lambda_k, \pi_k, k \in \mathcal{N}$ are scalar parameters.

Through replacing the components (2.48) with (2.57), the matrix inequality (2.55) is turned into an LMI. Moreover, the minimization of the off-diagonal blocks P_{kj} ensures that they are only as large as needed. While setting $P_{kj} = 0$ is too restrictive as discussed in Section 2.2.3, large off-diagonal blocks P_{kj} inhibit the design of suitable observer gain matrices. Moreover, in Step 1, L_k is replaced by $\lambda_k C_k^\top$ and K_{kj} is replaced by $\lambda_k I$ in the terms $\widetilde{Q}_{k,j}$ and $\widetilde{S}^k_{j,i}$ for preliminary calculation. In the second step, the exact observer gains are subsequently calculated.

Step 2: Set $P_k, k \in \mathcal{N}$ and $P_{kj}, j \in \mathcal{N}_k$ as the results from Step 1. Let $\widetilde{Q}_{k,k} = Q_{k,k}, \widetilde{Q}_{k,j} = Q_{k,j}, \widetilde{S}^k_{j,i} = S^k_{j,i}$ be defined according to (2.48) and $\widetilde{T}_{k,j} = (B^\phi - L_k D_k)^\top P_{kj}$. Then, solve (2.55) subject to the remaining variables.

An example where this 2 step approach is used will be given in the next section. It is a heuristic approach, but has proven capable of solving cases where the intuitive approach from Section 3 fails. In particular, it is able to resolve the conservatism of the convex problem presented in Section 2.2.3. There is room for improving the heuristics, but this approach already shows how the task of solving the matrix inequalities (2.55) can be tackled.

Remark 2.8 *The performance of the resulting observers with respect to the exogenous input w is determined by choice of γ. Small γ therefore leads to large observer gains. The parameters λ_k determine the effects of the mixed terms $L_k P_{kj}$ and $K_{kj} P_{kj}$ on the first solution step. If λ_k is set small, it can cause step 2 to be infeasible.*

2.2.6 Simulation example

The example presented in the following demonstrates the limitations of the approach from Section 2.2.2 and shows the value of the extended observer design from Section 2.2.4. Let a six-dimensional oscillator given by

$$\dot{x} = \begin{bmatrix} 0 & 1 & 0 & 1 & 0 & 1 \\ -1 & 0 & 1 & 0 & 1 & 0 \\ 0 & -1 & 0 & 1 & 0 & 1 \\ -1 & 0 & -1 & 0 & 1 & 0 \\ 0 & -1 & 0 & -1 & 0 & 1 \\ -1 & 0 & -1 & 0 & -1 & 0 \end{bmatrix} x + \begin{bmatrix} 1 \\ 0 \\ 0 \\ -1 \\ 0 \\ 0 \end{bmatrix} \phi(Hx) + \begin{bmatrix} 0 \\ 0 \\ 0 \\ 0 \\ 0 \\ 1 \end{bmatrix} w$$

with the monotonously increasing nonlinearity $\phi(\cdot)$, which are set as

$$\phi(Hx) = \sqrt[3]{Hx}$$
$$H = \begin{bmatrix} 1 & 1 & 1 & 1 & 1 & 1 \end{bmatrix}$$

in the following simulations.

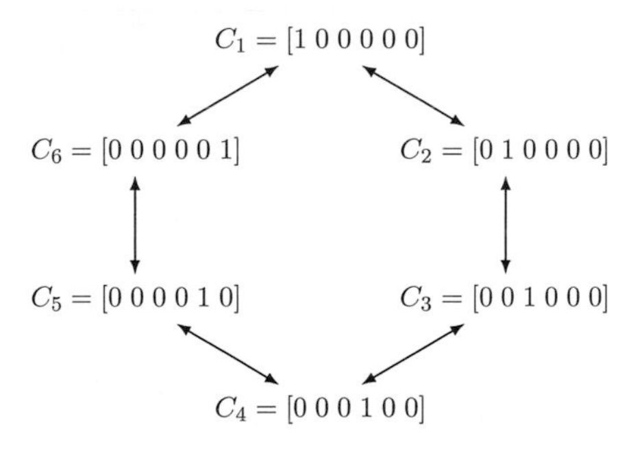

Figure 2.3: Communication topology of the observers.

Let the individual measurements be

$$y_k = x_k, \quad k \in \mathcal{N} \tag{2.58}$$

where $x = [x_1, ..., x_6]^\top$, and let the observers be connected by a ring-type communication topology shown in Figure 2.3.

The nonlinearity ϕ does not satisfy any global Lipschitz condition, but can be characterized using the incremental quadratic constraint (2.26), where $M = \begin{bmatrix} 0 & I \\ I & 0 \end{bmatrix}$. With this kind of multiplier matrix M, (2.39) is infeasible as discussed in Section 2.2.3. To be precise, (2.39) implies the equality

$$P_k B^\phi = -H^\top + C_k^\top \widetilde{L}_k^\top \tag{2.59}$$

for all $k \in \mathcal{N}$, which is infeasible.

In contrast, applying the 2-step approach from Section 2.2.4 relaxes above equality to

$$P_k B^\phi = -H^\top + C_k^\top \widetilde{L}_k^\top + \sum_{j \in \mathcal{N}_k} \widetilde{K}_{kj}^\top. \tag{2.60}$$

With the choice of parameters $\pi_k = 0.5$, $\lambda_k = 10$ for all $k \in \mathcal{N}$, and the performance parameter $\gamma = 2$ the procedure is feasible. Simulations of the system states x_1, x_2 are shown in Figure 2.4. In the time interval $[10, 20]$, the system is disturbed by $w(t) = 2$, which amplifies the oscillation. As it can be seen in Figure 2.5, the estimates converge towards the real states while there are no disturbances and during the time interval $[10, 20]$, the disturbance is attenuated.

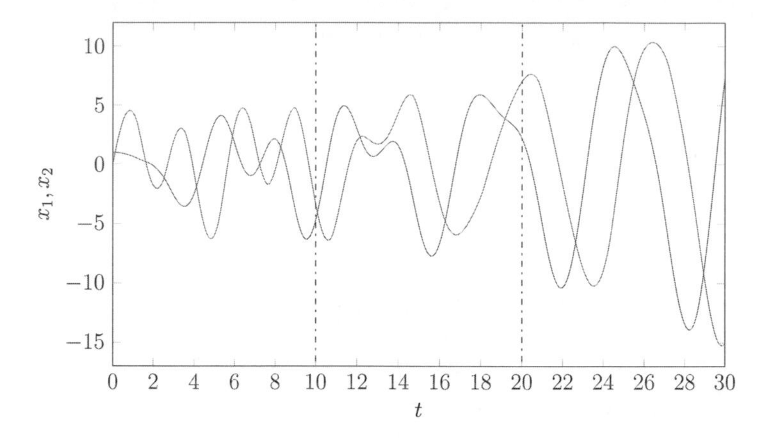

Figure 2.4: Plots of x_1 and x_2.

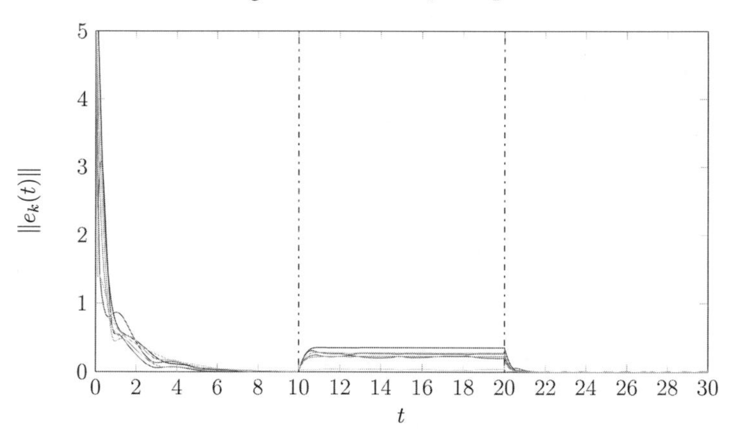

Figure 2.5: Plots of the estimation error of all observers.

2.3 Summary and Discussion

In this chapter, we presented distributed estimation schemes for both linear systems and systems with additive nonlinearity satisfying an incremental quadratic constraint. To the best of our knowledge, considering nonlinear systems for distributed estimation has not been be studied in the literature before and is important to make these methods relevant to a real-world applications.

The design conditions for the observer gain matrices were presented in the form of distributed LMIs, i.e. every observer is assigned an individual LMI condition, but the LMIs of neighbours posses coupling variables. This relates those LMI conditions to the existence of a vector storage function, and is a viable approach to ensure convergence and performance for the distributed estimation of linear systems.

In the nonlinear case however, assuming purely quadratic terms in the storage function candidate is shown to be highly restrictive. In order to demonstrated this, a close relation to the theory of absolute stability is established and the KYP-Lemma is used to show the properties of the resulting conditions. Therefore, an extended approach is proposed based on the inclusion of mixed-terms into the storage function candidate. Still, the design conditions can be presented in the form of distributed LMIs and simulations illustrate how the restrictions of the pure quadratic approach are lifted.

The class of nonlinearities which can be described using incremental quadratic constraints includes several important subclasses as discussed in this chapter. The subsequent step to extend the class of nonlinearities would be considering methods using multipliers (Zames & Falb, 1968; Materassi & Salapaka, 2011) or integral quadratic constraints (Jonsson & Megretski, 2000; Veenman & Scherer, 2011), which may yield considerable extension towards to class of nonlinearities that can be considered for distributed estimation.

Moreover, in this Chapter we have assumed communication channels to be perfect and transmit in continuous-time, which is a simplification from the practical perspective. In order to enable application, realistic modelling of the communication channels and associated disturbances is an important issue. In the following, we discuss three ways to take communication channels into account and how to address the resulting issues:

1. If communication is implemented with analogue signals, communication errors may be modelled as additive noise. Such kind of disturbances can be included into the $\mathcal{H}_\infty$-type design framework and the LMI-conditions, where some work in this direction has been done in (Zamani & Ugrinovskii, 2014).

2. If communication is implemented digitally, the communication channels may be subject to delays and package drop-outs. In recent years, many results have been published taking such disturbances into account. The standard problem setup is concerned with one plant and one controller, where delays or drop-outs happen between these two, see e.g. (Zhang et al., 2001; Schenato et al., 2007). Respectively, when only the estimation problem is considered, then disturbances are considered in the measurement channel, see e.g. (Fattouh et al., 1998; Fridman &

Shaked, 2001; Schenato, 2008; Briat et al., 2011). To the best of our knowledge, in distributed estimation schemes, delays and packet-dropouts in the communication channels between the observers have not yet been studied. However, in the case of our proposed distributed observers (2.6), we expect that delays can be incorporated by applying a Lyapunov-Kravoskii-function similar to (Fridman & Shaked, 2001), which will lead to LMI-conditions that are sufficient to solve Problem 1 despite the communication delays. However, the resulting LMI-conditions will further grow in complexity.

3. Reducing the frequency of communication is another issue, which has attracted a lot of attention in control theory lately. In order to save bandwith and energy, event-triggered communication strategies can be applied to our distributed estimation scheme. This is an important aspect to enhance practicability and will therefore be elaborated in the next Chapter.

Lastly, solving the LMI condition presented in this chapter is another important issue that needs to be thoroughly discussed. In particular, since we established the LMI-conditions in a distributed form, it is meaningful to enable solution of these conditions in a distributed fashion. Besides event-triggered communication, this will also be a subject of the next Chapter.

Chapter 3

Event-triggered Communication and Distributed Design

In the previous chapter, we introduced a distributed observer scheme where individual observers can cooperatively estimate the underlying system's state without a central coordination unit. However, as discussed in Section 2.3, more effort needs to be spend in order to make this scheme practicable, where in the following, we will focus on two aspects, namely reducing transmission by applying event-triggered communication, and enabling distributed computation of the observer gain matrices. These two extensions will be considered for the linear estimation problem presented in Section 2.1, and they will be presented separately. However, combinations of these two extensions and the nonlinear estimation scheme from Section 2.2 may also be considered.

Event-triggered control was introduced in (Bernhardsson & Åström, 1999) and has since then drawn a lot of attention, where some important reference, among many others, are (Tabuada, 2007; Mazo & Tabuada, 2011; Dimarogonas & Johansson, 2009; Dimarogonas et al., 2012; Seyboth et al., 2013). The latter two references thoroughly discuss consensus problems of multi-agent systems, where communication between the agents is subject to triggering. The goal of event-triggering in general is to transmit information whenever it is valuable, instead of transmitting information continuously. Therefore, considering event-triggered communication in distributed estimation schemes such as (2.6) is an important extension that significantly enhances applicability. Some related results have been published in (Shi et al., 2014; Ge & Han, 2015), where the transmission of measurements from the sensors to the observers is triggered. In contrast, we aim at event-based transmission of the estimates between the observers here, which has not been considered in the literature yet to the best of our knowledge.

Furthermore, practicality of a distributed system demands that the estimator design process is to be carried out in a distributed manner as well. This allows each observer to reconfigure using local communication and computation only, when the observers need to adapt to some changes, such as a change in the plant or change in the network structure.

This Chapter is structured as follows: In Section 3.1, we present a modified version of the distributed observers (2.6), where the observers do not receive their neighbours' estimate continuously, but only when a trigger rule is active. We prove that the trigger rule is well behaved in the sense that there is a lower bound on the time intervals between trigger instances (*inter-event times*). In Section 3.2, we present an approach to enable distributed computation of the observer gain matrices, which directly builds upon the methodology of distributed LMI-conditions. The approach is based on well-known technique from distributed optimization, where the coupling variables are separated and the optimization problem is reformulated using dual decomposition methods. This section is based on (Wu, Li et al., 2015). Lastly, in Section 3.3, we give an overview on our recent findings based on a *Distributed Kalman Filtering*. In this approach, every observer implements a Riccati Differential Equation (RDE) whose solution is used to adapt the observer gain matrices. The special feature here lies in the usage of couplings between the RDEs in order to ensure that the solution remains bounded and thus, the observer gain matrices remain bounded, unlike (Olfati-Saber, 2007), where the estimates are coupled, but the RDEs are not. This Section is based on (Wu, Elser et al., 2016) and also refers to (J. Kim, Shim & Wu, 2016).

3.1 Event-triggered Communication

The event-triggered control is based on the idea that information transmission is triggered, whenever some error function ϵ exceeds a certain threshold. This becomes valuable in cases where the effort for monitoring ϵ is tolerable, but communication is expensive due to limited bandwidth or limited energy resources. For instance, if a controller receives sensor measurements through some communication network, then it is meaningful that new measurement data is only transmitted when the measurement actually changes. Transferring this idea to our distributed estimation scheme from Chapter 2, we seek for a ways to schedule communication between the individual observers (2.6). In particular, it is important to find local trigger rules, i.e. local variables ϵ_k that can be monitored without communication.

3.1.1 Observer scheme

We consider the linear system (2.1) with the measurements (2.2) and modify the distributed observers (2.6) in the following way:

$$\dot{\hat{x}}_k = A\hat{x}_k + L_k(y_k - C_k\hat{x}_k) + K_k \sum_{j\in\mathcal{N}_k} (\widetilde{x}_j - \hat{x}_k) \tag{3.1}$$

where $\widetilde{x}_j$ is the projected estimate of $\hat{x}_j$. Let an event be triggered for observer k at time t_k^i. Then, at time t_k^i, observer k broadcasts its estimate $\hat{x}_k$ to its outgoing neighbourhood $\mathcal{M}_k$, where we assume here that there are no packet losses. Hence, at t_k^i, $\widetilde{x}_k(t_k^i) = \hat{x}_k(t_k^i)$.

In between the trigger instances, i.e. during the intervals $[t_k^i, t_k^{i+1})$, the dynamics of $\widetilde{x}_k$ are given as

$$\dot{\widetilde{x}}_k = A\widetilde{x}_k. \tag{3.2}$$

Note that the projected estimate $\widetilde{x}_k$ is implemented by observer k and all its outgoing neighbours $j \in \mathcal{M}_k$, so to be precise, there are $q_k + 1$ realisations of $\widetilde{x}_k$, which all have the same value. In between the trigger instances, no communication is needed to update $\widetilde{x}_k$. Moreover, observer k is able to monitor its *trigger error*, which is given as $\epsilon_k = \hat{x}_k - \widetilde{x}_k$. The trigger function is therefore proposed as

$$h_k(t, \epsilon_k) = \|\epsilon_k(t)\|^2 - (c_0 + c_1 \exp(-\tilde{\alpha}t)) \tag{3.3}$$

with $c_0, c_1 \geq 0$, $c_0 + c_1 > 0$, and $\tilde{\alpha} > 0$. Events are thus triggered for observer k, when $h_k(t, \epsilon_k) = 0$ and then, ϵ_k is reset to 0.

Remark 3.1 *It should be emphasized that in most event-triggered schemes, the transmitted variable $\widetilde{x}_k$ is subject to sample-hold, i.e. $\widetilde{x}_k(t) = \hat{x}_k(t_k^i)$ for all $t \in [t_k^i, t_k^{i+1})$. In contrast, in our scheme, $\widetilde{x}_k$ is propagated with* (3.2), *which is used because this way, we can achieve that in steady state with $\hat{x}_k \equiv x, k \in \mathcal{N}$, it holds that $\widetilde{x}_k \equiv \hat{x}_k$.*

Note that when $c_0 > 0$, then exact convergence of the estimation errors $e_k = x - \hat{x}_k$ cannot be guaranteed even for the nominal case with $w \equiv 0$ and $\eta_k \equiv 0$. On the other hand, when $c_0 = 0$, then agglomeration of trigger instances cannot be excluded for the disturbed case with $w, \eta_k \neq 0$. Therefore, with the modified distributed observers, the problem statement also needs to be adapted.

Problem 3 (Event-triggered distributed observers) *For all $k \in \mathcal{N}$, determine observer gain matrices L_k and K_k for* (3.1), *such that*

1. *In the nominal case, i.e.* $w \equiv 0, \eta_k \equiv 0$ *for all* $k \in \mathcal{N}$*, we have convergence of the estimates to a ball centered around the origin, in the sense that*

$$\lim_{t\to\infty} \|\hat{x}_k(t) - x(t)\| \le \theta c_0, \tag{3.4}$$

with $\theta > 0$ *and for all initial conditions* $\hat{x}_k(0), x(0)$.

2. *For a given positive semi-definite weighting matrix* W_k *we have performance of the estimation errors* $e_k = x - \hat{x}_k$ *in the sense that*

$$\limsup_{T\to\infty} \left(\frac{1}{T} \sum_{k=1}^{N} \int_0^T e_k^\top W_k e_k \right) \le \limsup_{T\to\infty} \left(\frac{\gamma^2}{T} \sum_{k=1}^{N} \int_0^T (\|w\|^2 + \|\eta_k\|^2) dt + I_0 \right), \tag{3.5}$$

where I_0 *is the cost due to the relaxed convergence properties with* $c_0 > 0$.

Property 2 represents an energy-type performance inequality, where the estimation errors e_k are required to be bounded by the disturbances, but some persistent deviation is acceptable. One more property, which needs to be addressed when designing the observers is the minimum inter-event time $\min_{k,i} |t_k^i - t_k^{i+1}|$, which needs to be strictly positive. Here, we aim at obtaining a direct relationship of the minimum inter-event time to the persistent error I_0, the disturbances w, η_k, and the parameter c_0.

3.1.2 Observer design conditions

The observer design can be performed similarly to the procedure in Section 2.1.3, which is presented in the following Theorem.

Theorem 3.1 *Let a collection of matrices* F_k, $\widetilde{P}_k \succ 0$, G_k *and* $P_k \succ 0$, $k \in \mathcal{N}$*, be a solution of the LMIs*

$$\left[\begin{array}{ccc:c} Q_k + W_k + q_k(1+\kappa)\widetilde{P}_k & -G_k & P_k B^w & \mathbf{1}_{p_k}^\top \otimes F_k \\ -G_k^\top & -\gamma^2 I & 0 & 0 \\ B^{w\top} P_k^\top & 0 & -\gamma^2 I & 0 \\ \hdashline \mathbf{1}_{p_k} \otimes F_k^\top & 0 & 0 & -diag\left[\widetilde{P}_j\right]_{j\in\mathcal{N}_k} \end{array}\right] \preceq 0 \tag{3.6}$$

$$Q_k = P_k A + A^\top P_k - G_k C_k - C_k^\top G_k^\top - p_k F_k - p_k F_k^\top + \alpha P_k$$

for all $k \in \mathcal{N}$*, where* $\alpha > 0$ *and* $\kappa > 0$*. Then, the observers* (3.1), (3.2) *with the observer gain matrices*

$$L_k = P_k^{-1} G_k$$
$$K_k = P_k^{-1} F_k.$$

and the trigger rule (3.3) *solve Problem 3.*

For the sake of simplicity, we assume here that α and the parameters of (3.3) are chosen identically for all observers.

Proof. Let the estimation error be given as $e_k = x - \hat{x}_k$, then, the derivative is

$$\begin{aligned}\dot{e}_k &= Ax + B^w w - A\hat{x}_k - L_k(y_k - C_k\hat{x}_k) - K_k \sum_{j\in\mathcal{N}_k} (\widetilde{x}_j - \hat{x}_k) \\ &= (A - L_kC_k)e_k - L_k\eta_k + B^w w + K_k \sum_{j\in\mathcal{N}_k} (\epsilon_j + e_j - e_k).\end{aligned} \tag{3.7}$$

With the storage function candidate component $V_k = e_k^\top P_k e_k$ we have the Lie-derivative

$$\begin{aligned}\dot{V}_k =& e_k^\top (P_kA + A^\top P_k - G_kC_k - C_k^\top G_k^\top - p_kF_k - p_kF_k^\top)e_k - 2e_k^\top F_k\eta_k + 2e_k^\top P_kB^w w \\ &+ 2e_k^\top F_k \sum_{j\in\mathcal{N}_k} (\epsilon_j + e_j)\end{aligned}$$

Now, suppose the LMIs (3.6) holds. Then, we have the dissipation inequality

$$\begin{aligned}\dot{V}_k =& - e_k^\top (q_k(1+\kappa)\widetilde{P}_k + \alpha P_k + W_k)e_k + \gamma^2\|\eta_k\|^2 + \gamma^2\|w\|^2 + \sum_{j\in\mathcal{N}_k} (\epsilon_j + e_j)^\top \widetilde{P}_j(\epsilon_j + e_j) \\ \leq& - e_k^\top (q_k(1+\kappa)\widetilde{P}_k + \alpha P_k + W_k)e_k + \gamma^2\|\eta_k\|^2 + \gamma^2\|w\|^2 \\ &+ \sum_{j\in\mathcal{N}_k} \left((1 + \frac{1}{\kappa})\epsilon_j^\top \widetilde{P}_j\epsilon_j + (1+\kappa)e_j^\top \widetilde{P}_j e_j\right).\end{aligned}$$

For the storage function $V = \sum_{k\in\mathcal{N}} V_k$, we have

$$\dot{V} \leq - \sum_{k\in\mathcal{N}} e_k^\top (W_k + \alpha P_k)e_k + \gamma^2 \sum_{k\in\mathcal{N}} (\|\eta_k\|^2 + \|w\|^2) + \sum_{k\in\mathcal{N}} q_k(1 + \frac{1}{\kappa})\epsilon_k^\top \widetilde{P}_k\epsilon_k \tag{3.8}$$

Now, in the nominal case with $w \equiv 0, \eta_k \equiv 0$, the dissipation inequality (3.8) can be simplified to

$$\begin{aligned}\dot{V} \leq& - \alpha \sum_{k\in\mathcal{N}} e_k^\top P_k e_k + \sum_{k\in\mathcal{N}} q_k(1 + \frac{1}{\kappa})\epsilon_k^\top \widetilde{P}_k\epsilon_k \\ \dot{V} \leq& - \alpha V + \sum_{k\in\mathcal{N}} q_k(1 + \frac{1}{\kappa})\|\epsilon_k\|^2_{\widetilde{P}_k}.\end{aligned}$$

With the triggering rule (3.3), we have

$$\begin{aligned}\|\epsilon_k\|^2 &\leq c_1\exp(-\widetilde{\alpha}t) + c_0 \\ \|\epsilon_k\|^2_{\widetilde{P}_k} &\leq \lambda_{max}(\widetilde{P}_k)(c_1\exp(-\widetilde{\alpha}t) + c_0)\end{aligned} \tag{3.9}$$

and therefore

$$\dot{V} \leq -\alpha V + \sum_{k\in\mathcal{N}} q_k(1+\frac{1}{\kappa})\lambda_{max}(\widetilde{P}_k)(c_1\exp(-\widetilde{\alpha}t)+c_0). \tag{3.10}$$

With the notation $\mu = N\max_k(q_k)(1+\frac{1}{\kappa})\max_k(\lambda_{max}(\widetilde{P}_k))$ and assuming for simplicity that $\alpha \neq \widetilde{\alpha}$, we have for the evolution of $V(t)$

$$\begin{aligned}
V(t) \leq & \exp(-\alpha t)V(0) + \mu\int_0^t \exp(-\alpha(t-\tau))(c_1\exp(-\widetilde{\alpha}\tau)+c_0)d\tau \\
\leq & \exp(-\alpha t)V(0) + \mu c_1\int_0^t \exp(-\alpha(t-\tau)-\widetilde{\alpha}\tau)d\tau + \mu c_0\int_0^t \exp(-\alpha(t-\tau))d\tau \\
\leq & \exp(-\alpha t)V(0) + \mu c_1\exp(-\alpha t)\frac{1}{\alpha-\widetilde{\alpha}}\left[\exp((\alpha-\widetilde{\alpha})t)-1\right] + \mu c_0\frac{1}{\alpha}\left[1-\exp(-\alpha t)\right] \\
\leq & \exp(-\alpha t)V(0) + \frac{\mu c_1(\exp(-\widetilde{\alpha}t)-\exp(-\alpha t))}{\alpha-\widetilde{\alpha}} - \frac{\mu c_0\exp(-\alpha t)}{\alpha} + \frac{\mu c_0}{\alpha}.
\end{aligned}$$

Since $\alpha > \widetilde{\alpha}$, we conclude that

$$\lim_{t\to\infty} V(t) \leq \frac{\mu c_0}{\alpha}.$$

Then, as $e_k^\top P_k e_k \leq V$, Property 1 of Problem 3 holds with $\theta = \mu/(\alpha\lambda_{\min}P_k)$.

For the disturbed case, the dissipation inequality (3.8) after integration over $[0,T]$ yields

$$\begin{aligned}
V(T) \leq & -\sum_{k\in\mathcal{N}}\int_0^T e_k^\top(W_k+\alpha P_k)e_k dt + \gamma^2\sum_{k\in\mathcal{N}}\int_0^T \|\eta_k\|^2+\|w\|^2 dt \\
& + \sum_{k\in\mathcal{N}}\int_0^T q_k(1+\frac{1}{\kappa})\epsilon_k^\top\widetilde{P}_k\epsilon_k dt + V(0)
\end{aligned}$$

With (3.9) we have

$$\sum_{k\in\mathcal{N}}\int_0^T e_k^\top W_k e_k dt \leq \gamma^2\sum_{k\in\mathcal{N}}\int_0^T \|\eta_k\|^2+\|w\|^2 dt + \mu\sum_{k\in\mathcal{N}}\int_0^T (c_1\exp(-\widetilde{\alpha}t)+c_0)dt + V(0)$$

and finally, after division by T and letting $T\to\infty$, we have

$$\limsup_{T\to\infty}\left(\frac{1}{T}\sum_{k=1}^N\int_0^T e_k^\top W_k e_k\right) \leq \limsup_{T\to\infty}\left(\frac{\gamma^2}{T}\sum_{k=1}^N\int_0^T (\|w\|^2+\|\eta_k\|^2)dt + I_0\right)$$

with $I_0 = \mu c_0$. ■

Remark 3.2 *The calculation of θ for Property 1 of Problem 3 is conservative and can be improved by eliminating conservative bounds. For instance, using the weighted squared norm $\|\epsilon_k\|^2_{\widetilde{P}_k}$ for the trigger function* (3.3) *instead of $\|\epsilon_k\|^2$ reduces conservatism.*

Theorem 3.1 shows that solving the LMI-conditions (3.6) is a suitable design method for obtaining the observer gain matrices L_k, K_k, that allow for application of the event-triggered communication scheme (3.1), (3.2), (3.3). However, as already indicated above, this result is not yet complete as agglomeration of trigger events, so-called *Zeno-behaviour* needs to be excluded. This will be addressed in the following.

3.1.3 Exclusion of Zeno-behaviour

By design, at a trigger instance t_k^i, ϵ_k is reset to 0. Thus, we aim at finding an uniform lower bound for the time that it takes for ϵ_k to trigger an event according to the trigger rule (3.3). For this purpose, we introduce the following mild assumption on the disturbances w, η_k.

Assumption 3.1 *The norm of the disturbances are uniformly upper bounded in the sense that $\|w(t)\|^2 \leq \overline{w}$, $\|\eta_k(t)\|^2 \leq \overline{\eta}$, $k \in \mathcal{N}$ for every time instance $t \geq 0$.*

Also, we assume that $c_0 > 0$, as this is required in the perturbed case in order to determine a lower bound on the inter-event time. The nominal case, with $c_0 = 0$ can however be considered analogously. For the trigger error ϵ_k, the derivative is

$$\dot{\epsilon}_k = A\epsilon_k + L_k C_k e_k + L_k \eta_k - K_k \sum_{j\in\mathcal{N}_k} (\epsilon_j + e_j - e_k) \tag{3.11}$$

during the intervals $[t_k^i, t_k^{i+1})$. Again, note that $\epsilon(t_k^i) = 0$. Then, it holds for $t \in [t_k^i, t_k^{i+1})$ that

$$\epsilon_k(t) = \int_{t_k^i}^{t} \exp(A(t-\tau)) B_k u_k(\tau) d\tau \tag{3.12}$$

where

$$u_k = \begin{bmatrix} e_k \\ [e_j]_{j\in\mathcal{N}_k} \\ [\epsilon_j]_{j\in\mathcal{N}_k} \\ \eta_k \end{bmatrix}, \qquad B_k = \begin{bmatrix} L_k C_k + p_k K_k & -\mathbf{1}_{2p_k}^\top \otimes K_k & L_k \end{bmatrix} \tag{3.13}$$

Then, the norm of $\epsilon_k(t)$ can be upper bounded by

$$\begin{aligned}\|\epsilon_k(t)\| &= \|\int_{t_k^i}^{t} \exp(A(t-\tau))B_k u_k(\tau)d\tau\| \\ &\leq \int_{t_k^i}^{t} \exp(\lambda_{max}(A)(t-\tau))\|B_k\|\|u_k(\tau)\|d\tau, \end{aligned} \tag{3.14}$$

where $\|B_k\|$ is the induced norm of B_k and $\|u_k(t)\|$ can be upper bounded by

$$\begin{aligned}\|u_k(t)\|^2 &\leq \|e(t)\|^2 + \sum_{j\in\mathcal{N}_k} \|\epsilon_j(t)\|^2 + \|\eta_k(t)\|^2 \\ &\leq \|e(t)\|^2 + p_k(c_1\exp(-\widetilde{\alpha}t) + c_0) + \overline{\eta} \end{aligned} \tag{3.15}$$

With Assumption 3.1, the dissipation inequality (3.8) implies

$$\dot{V} \leq -\alpha V + \mu(c_1\exp(-\widetilde{\alpha}t) + c_0) + N\gamma^2(\overline{w} + \overline{\eta}),$$

where again, $\mu = N\max_k(q_k)(1+\frac{1}{\kappa})\max_k(\lambda_{max}(\widetilde{P}_k))$. By redoing the calculations from the proof of Theorem 3.1, we arrive at

$$\begin{aligned}V(t) \leq& \exp(-\alpha t)V(0) + \frac{\mu c_1(\exp(-\widetilde{\alpha}t) - \exp(-\alpha t))}{\alpha - \widetilde{\alpha}} \\ &- \frac{(\mu c_0 + N\gamma^2(\overline{w}+\overline{\eta}))\exp(-\alpha t)}{\alpha} + \frac{\mu c_0 + N\gamma^2(\overline{w}+\overline{\eta})}{\alpha} \\ \min_k(\lambda_{min}(P_k))\|e(t)\|^2 \leq& \exp(-\alpha t)\left(V(0) - \frac{\mu c_1}{\alpha - \widetilde{\alpha}} - \frac{(\mu c_0 + N\gamma^2(\overline{w}+\overline{\eta}))}{\alpha}\right) \\ &+ \exp(-\widetilde{\alpha}t)\frac{\mu c_1}{\alpha-\widetilde{\alpha}} + \frac{\mu c_0 + N\gamma^2(\overline{w}+\overline{\eta})}{\alpha} \end{aligned} \tag{3.16}$$

With (3.16) it can be seen that the right-hand-side of (3.15) is monotonically decreasing. Therefore, for every $k \in \mathcal{N}$ there exists a constant bound $u_{\max,k}$ such that $\|u_k(t)\| \leq u_{\max,k}$. Then, with (3.14), we have

$$\begin{aligned}\|\epsilon_k(t)\| &\leq \|B_k\|u_{\max,k}\int_{t_k^i}^{t} \exp(\lambda_{max}(A)(t-\tau))d\tau \\ &= \|B_k\|u_{\max,k}\frac{\exp(\lambda_{max}(A)(t-t_k^i)) - 1}{\lambda_{max}(A)}\end{aligned}$$

For the tigger instance t_k^{i+1}, it holds that $\|\epsilon_k(t_k^{i+1})\| \geq c_0$, which implies that

$$\begin{aligned}\|B_k\|u_{\max,k}\frac{\exp(\lambda_{max}(A)(t_k^{i+1}-t_k^i)) - 1}{\lambda_{max}(A)} &\geq c_0 \\ t_k^{i+1} - t_k^i &\geq \frac{1}{\lambda_{max}(A)}\log\left(\frac{c_0\lambda_{max}(A)}{\|B_k\|u_{\max,k}} + 1\right) > 0.\end{aligned} \tag{3.17}$$

For above lower bound on the inter-event time, we use a series of conservative approximations. It is sufficient to exclude Zeno behaviour, but more interestingly, we can calculate a lower bound on the persistent inter-event time in a less conservative way. From (3.16), we know that

$$\limsup_{t\to\infty} \|e(t)\|^2 \leq \frac{\mu c_0 + N\gamma^2(\overline{w}+\overline{\eta})}{\alpha \min_k(\lambda_{min}(P_k))} \tag{3.18}$$

and then, with (3.15), we have

$$\limsup_{t\to\infty} \|u_k(t)\|^2 \leq \frac{\mu c_0 + N\gamma^2(\overline{w}+\overline{\eta})}{\alpha \min_k(\lambda_{min}(P_k))} + p_k c_0 + \overline{\eta}. \tag{3.19}$$

For obtaining the persistent lower bound on the inter-event time, it suffices to replace $u_{\max,k}$ with the right-hand-side of (3.19). Hence, the lower bound obtained this way is influenced by the system parameters in the following way:

Increasing inter-event time	α	$\min_k(\lambda_{min}(P_k))$	c_0
Decreasing inter-event time	γ	$\overline{w}, \overline{\eta}$	L_k, K_k

Remark 3.3 *It can also be shown that in the undisturbed case with $w \equiv 0, \eta_k \equiv 0$, and the choice of parameters $c_0 = 0, c_1 > 0, W_k = 0, \alpha > \tilde{\alpha}$, the estimation errors e_k converge to 0, while Zeno behaviour can be excluded.*

3.1.4 Simulation example

Similar to the example of the previous Chapter, let a six-dimensional oscillator be given by

$$\dot{x} = \begin{bmatrix} 0 & 1 & 0 & 1 & 0 & 1 \\ -1 & 0 & 1 & 0 & 1 & 0 \\ 0 & -1 & 0 & 1 & 0 & 1 \\ -1 & 0 & -1 & 0 & 1 & 0 \\ 0 & -1 & 0 & -1 & 0 & 1 \\ -1 & 0 & -1 & 0 & -1 & 0 \end{bmatrix} x + \begin{bmatrix} 0 \\ 0 \\ 0 \\ 0 \\ 0 \\ 1 \end{bmatrix} w,$$

let the individual measurements again be

$$y_k = x_k, \quad k \in \mathcal{N}$$

where $x = [x_1, ..., x_6]^\top$, and let the observers be connected by a ring-type communication topology shown in Figure 3.1.

We choose the parameters $\pi_k = 0.1$, for all $k \in \mathcal{N}$, and minimize the performance parameter γ, where we achieve a value of 4.3.

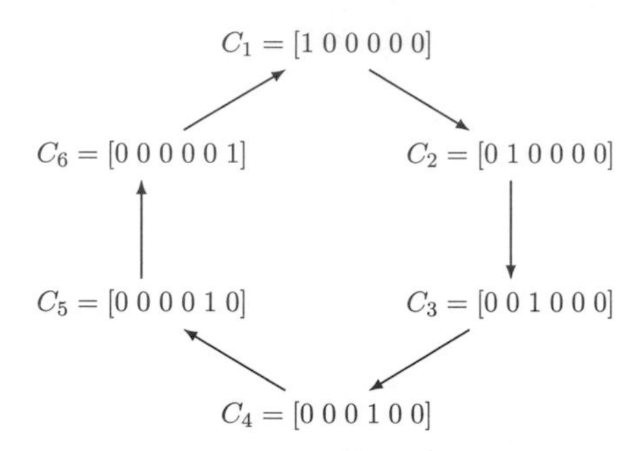

Figure 3.1: Communication topology of the observers.

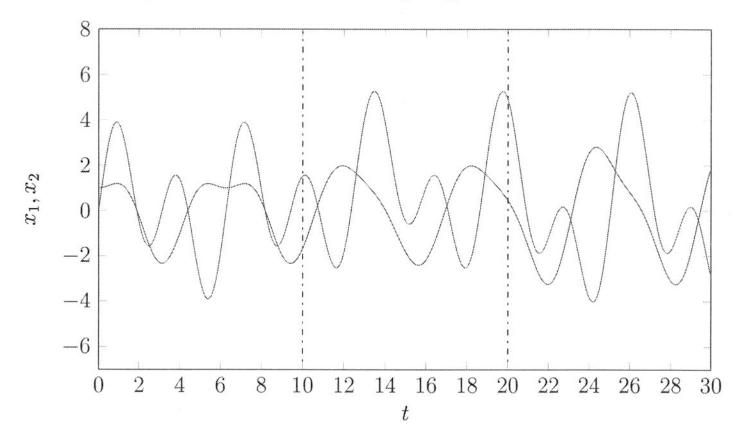

Figure 3.2: Plots of x_1 and x_2.

Simulations of the system states x_1, x_2 are shown in Figure 3.2 with the trigger parameters $c_0 = 1$ and $c_1 = 5$. In the time interval $[10, 20]$, the system is disturbed by $w(t) = 2$ and white noise $\eta_k(t)$. As it can be seen in Figure 3.3, the estimates converge towards the real states while there are no disturbances and during the time interval $[10, 20]$, the disturbance is attenuated. Figure 3.4 shows the evolution of the monitored error $\epsilon_1(t)$ and the triggering threshold. It can be seen that during undisturbed operation, only few transmissions are required, while in order to achieve disturbance attenuation, transmissions occur more frequently.

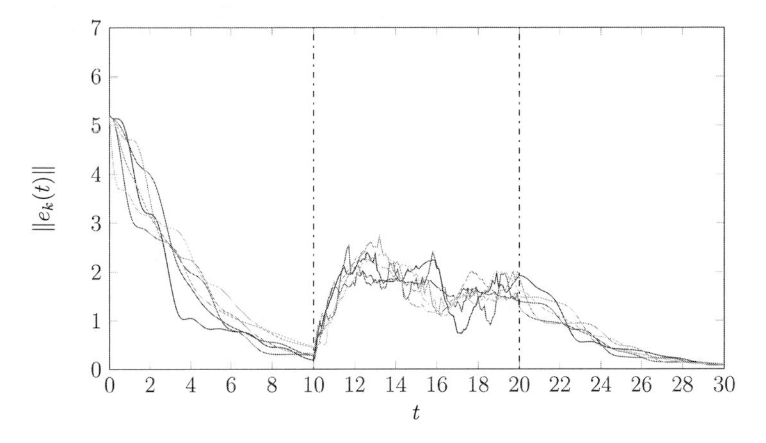

Figure 3.3: Plots of the estimation error of all observers.

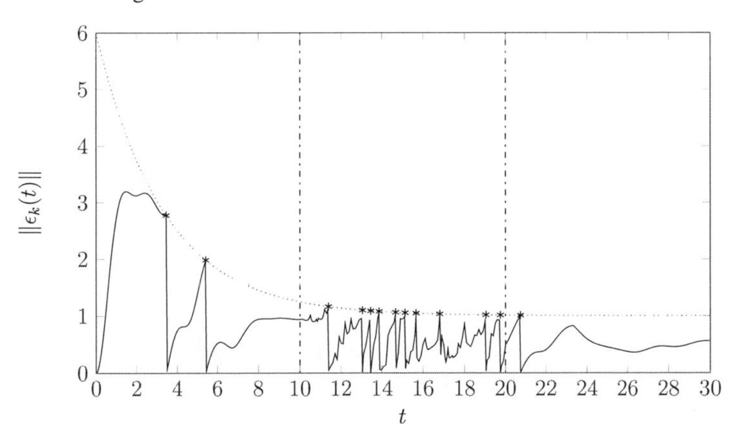

Figure 3.4: Monitored error ϵ_1. Threshold exponentially decays to $c_0 = 1$.

3.2 Dual Decomposition and Iterative optimization

The design options that we derived in Chapter 2 either consist of one global LMI which needs the knowledge of all measurement outputs y_k, $k \in \mathcal{N}$, or N LMIs which are coupled through matrix variables that are common to neighbouring agents. In fact, considering the literature review on distributed control in the introduction of this thesis, one notices that many publications present schemes of distributed controllers where design process itself requires a central coordination unit. If using a central coordination unit for the design procedure is acceptable, both our options for distributed estimation can be solved using conventional optimization tool such as YALMIP (Löfberg, 2004). In some practical application examples, where the design process can be done offline, this may not be a significant drawback. On the other hand, in many applications especially those involving distributed sensor networks with varying topology, a centralized computation of observer parameters represents a severe limitation. In this section, we present a method, which directly build upon the observer design conditions from Chapter 2, but enables solution in a distributed fashion, such that the group of observers can adapt to changes by interaction through their communication topology.

We make the following simplifying assumptions:

Assumption 3.2 *The communication graph $\mathcal{G}$ is connected and bi-directional.*

Assumption 3.3 *The pair (A, B^w) is controllable.*

Assumption 3.2 is a restriction on the class of communication graphs, which is made in order to simplify calculations. This assumption can be relaxed, as will be discussed later. Assumption 3.3 is used to ensure boundedness of the feasible sets. It is not restrictive, as it represents the worst case of disturbance, and if not satisfied, small fictitious disturbances can be added to the system description, i.e. additional columns to B^w, until Assumption 3.3 is satisfied.

3.2.1 Separation of the optimization problem

We revisit the LMI conditions (2.21) for designing the distributed observers (2.6), and reduce the number of variables by defining $\widetilde{P}_k = \pi_k P_k$ and $G_k = \psi_k C_k^\top$. This reduction will later be important to restrict the solution to a bounded set, and thus, make the optimization steps attainable. Furthermore, the first reduction is thoroughly discussed in Section 2.1.4. In addition, γ^2 is substituted by β in order to optimize the performance, which is Property 2 of Problem 1. Thus, the LMI conditions read

$$\left[\begin{array}{ccc:c} Q_k + W_k & -\psi_k C_k^\top & P_k B^w & \mathbf{1}_{p_k}^\top \otimes F_k \\ -\psi_k C_k & -\beta I & 0 & 0 \\ B^{w\top} P_k & 0 & -\beta I & 0 \\ \hdashline \mathbf{1}_{p_k} \otimes F_k^\top & 0 & 0 & -\mathrm{diag}\,[\pi_j P_j]_{j\in\mathcal{N}_k} \end{array}\right] \preceq 0 \tag{3.20}$$

$$Q_k = P_k A + A^\top P_k - 2\psi_k C_k^\top C_k - p_k F_k - p_k F_k^\top + (\alpha_k + p_k \pi_k) P_k.$$

for all $k \in \mathcal{N}$, where $P_k^k \succ 0 \in \mathbb{R}^{n_k \times n_k}$ and $F^k \in \mathbb{R}^{n_k \times n_k}$ and $\beta > 0$. Clearly, if (3.20) admits a solution, then so does (2.21) and the observers (2.6) with the observer gain matrices

$$L_k = \psi_k P_k^{-1} C_k$$
$$K_k = P_k^{-1} F_k$$

solve Problem 1.

Parallel and distributed computation is thoroughly discussed e.g. in (Bertsekas & Tsitsiklis, 1989). The technique of separating the variables is generally formulated as follows: Consider the optimization problem

$$\begin{aligned} &\min_{Y} \sum_{k=1}^{N} F_k(Y) \\ &\text{subject to } \; Y \in \Omega_k, \quad k \in \mathcal{N}, \end{aligned}$$

where Ω_k are some closed, convex sets. Then, the equivalent, separable version can be written as

$$\begin{aligned} &\min_{Y_k} \sum_{k=1}^{N} F_k(Y_k), \quad k \in \mathcal{N} \\ &\text{subject to } \; Y_k = Y, \quad k \in \mathcal{N}, \\ &\qquad\qquad Y_k \in \Omega_k, \quad k \in \mathcal{N}, \end{aligned} \tag{3.21}$$

where Y_k are artificial representations of the solution variable Y. In order to address the equality constraints $Y_k = Y$ involved in the optimization problem, the method of multipliers can be applied. Some references on this technique, among many others, are (Bertsekas & Tsitsiklis, 1989; Boyd et al., 2011), and more applications, where distributed optimization methods are illustrated can be found in (Raffard et al., 2004; Rabbat & Nowak, 2004; Necoara et al., 2011).

Our optimization problem involving the LMIs (3.20) can be cast in the sense of (3.21) by introducing local representations of the solution variables, P_j^k, and β^k for all $k \in \mathcal{N}$ and $j \in \mathcal{N}_k$. The tuple of local variables is denoted by

$$Y_k = (F_k, \beta^k, P_k^k, [P_j^k]_{j \in \mathcal{N}_k}), \tag{3.22}$$

where the upper index k denotes the representation of a variable used by observer k and all P_j^k are symmetric, positive definite matrices, and $\beta^k \geq 0$.

Thus, solving (3.20) and thereby minimizing β can be formulated as the separable problem

$$\text{minimize} \sum_{k=1}^{N} \beta_k \tag{3.23}$$

$$\text{subject to} \left[\begin{array}{ccc:c} Q_k + W_k & -\psi_k C_k^\top & P_k^k B^w & \mathbf{1}_{p_k}^\top \otimes F_k \\ -\psi_k C_k & -\beta_k I & 0 & 0 \\ B^{w\top} P_k^k & 0 & -\beta_k I & 0 \\ \hdashline \mathbf{1}_{p_k} \otimes F_k^\top & 0 & 0 & -\text{diag}\left[\pi_j P_j^k\right]_{j\in\mathcal{N}_k} \end{array}\right] \preceq 0, \quad k \in \mathcal{N}, \tag{3.24}$$

$$\beta^k = \widetilde{\beta}, \qquad k \in \mathcal{N}, \tag{3.25}$$

$$P_j^k = \widetilde{P}_j, \qquad k \in \mathcal{N}, j \in \mathcal{N}_k, \tag{3.26}$$

for some slack variables $\widetilde{\beta} \geq 0$ and $\widetilde{P}_j \succ 0$ and with

$$Q_k = P_k^k A + A^\top P_k^k - 2\psi_k C_k^\top C_k - p_k F_k - p_k F_k^\top + (\alpha_k + p_k \pi_k) P_k^k. \tag{3.27}$$

The matrices $P_j^k \succ 0$ and $\beta^k \geq 0$ are introduced here as local variables that replace P_j and β in the LMIs (3.20). Unlike (3.20), the LMIs (3.24) are decoupled and can by solved at observer k independently from the neighbours.

The Problem considered in this section thus consists of iteratively finding a solution to the optimization problem (3.23)-(3.26) in a distributed fashion.

One important property, which needs to be taken care of is that every iteration step is attainable. In the generic setup (3.21), it is often assumed in the literature that Ω_k is compact in order to guarantee that every optimization step is attainable, i.e. a minimum of $F_k(Y_k)$ can be found on Ω_k, see e.g. (Bertsekas & Tsitsiklis, 1989). However, note that the set of Y_k which solves (3.24) is not bounded. For this reason, for all $k \in \mathcal{N}$ we define the additional conditions

$$\begin{bmatrix} -\rho_k P_k^k & F_k^\top \\ F_k & -P_k^k \end{bmatrix} \preceq 0 \tag{3.28}$$

$$P_k^k \succeq \epsilon_k I, \qquad P_j^k \succeq \epsilon_k I, j \in \mathcal{N}_k, \tag{3.29}$$

with arbitrarily large $\rho_k > 0$ and arbitrarily small $\epsilon_k > 0$. The LMI (3.28) ensures that the feasible set of F^k is bounded through P_k^k and the LMIs (3.29) ensure that the feasible set is closed. The feasible sets Ω_k can now be defined as

$$\Omega_k = \{Y_k | \text{LMIs (3.24), (3.28), (3.29) are satisfied}\},$$

and will be used for the iterative optimization steps, shown in the next section. The set Ω_k is still unbounded because of the variables $P_j^k, j \in \mathcal{N}_k$. However, the respective optimizations on Ω_k are now attainable, which will be shown later. Note that ρ_k should be chosen large enough, such that the optimization (3.23), subject to $Y_k \in \Omega_k$ and the equality constraints (3.25), (3.26), is feasible.

Remark 3.4 *The optimization problem* (3.23) *can be varied in the way that for a given performance parameter* $\beta > 0$*, filter gains for* (2.6) *are to be found. Then,* (3.23)-(3.26) *turns to a pure feasibility problem without optimization objective, and therefore, the variables* β^k *and their iterations in the following algorithm can be omitted.*

The equality constraints (3.25) and (3.26) are incorporated into the objective function via the Lagrange multiplier technique, where we use the augmented Lagrangian

$$\begin{aligned} L = &\sum_{k=1}^{N} \left(\beta^k + \lambda^k (\widetilde{\beta} - \beta^k) + \frac{c}{2} |\widetilde{\beta} - \beta^k|^2 \right) \\ &+ \sum_{k=1}^{N} \sum_{j \in \overline{\mathcal{N}}_k} \left(tr\left(\Lambda_j^{k\top} (\widetilde{P}^j - P^{j|k}) \right) + \frac{c}{2} \| \widetilde{P}^j - P^{j|k} \|^2 \right) \end{aligned} \tag{3.30}$$

with $c > 0$ and the Frobenius-norm $\| \cdot \|$ in order to achieve strict convexity of the optimization objective functions. The tuple of local Lagrange multipliers is defined as $\widetilde{\Lambda}_k = (\lambda_k, \Lambda_k^k, [\Lambda_j^k]_{j \in \mathcal{N}_k})$. Using this method, the optimization problem (3.23) can now be cast into the unconstrained problem

$$\text{maximize } \mathbf{q}(\widetilde{\Lambda}_1, ..., \widetilde{\Lambda}_N), \tag{3.31}$$

where the dual function $\mathbf{q}(\cdot)$ is defined as

$$\mathbf{q}(\widetilde{\Lambda}_1, ..., \widetilde{\Lambda}_N) = \inf_{Y_k \in \Omega_k, k \in \mathcal{N}} L(Y_1, ..., Y_N, \widetilde{\Lambda}_1, ..., \widetilde{\Lambda}_N).$$

3.2.2 Iterative optimization algorithm

To initiate the iterative algorithm, initial conditions $Y_k(0) \in \Omega_k$ are computed by solving the LMIs (3.24), (3.28), (3.29) locally. Also, we set initial values for the Lagrangian

Algorithm 1 Iteration step t+1

1: Compute fusion variables for $j = 1, ..., N$,

$$\beta(t+1) = \frac{1}{N}\sum_{k=1}^{N}\beta^k(t) - \frac{1}{Nc}\sum_{k=1}^{N}\lambda^k(t)$$
$$P_j(t+1) = \frac{1}{q_j+1}\sum_{k\in\overline{\mathcal{N}}_j} P_j^k - \frac{1}{(q_j+1)c}\sum_{k\in\overline{\mathcal{N}}_j}\Lambda_j^k(t)$$

2: Update Y_k for all $k \in \mathcal{N}$,

$$Y_k(t+1) = \arg\min_{Y_k\in\Omega_k}\Bigg(\beta_k - \lambda_k(t)\beta_k + \frac{c}{2}|\beta(t+1) - \beta_k|^2 + \sum_{j\in\overline{\mathcal{N}}_k}\left(-\mathrm{tr}\left(\Lambda_j^{k\top}(t)P_j^k\right) + \frac{c}{2}\|P_j(t+1) - P_j^k\|^2\right)\Bigg)$$

3: Update the Lagrange multipliers for $k \in \mathcal{N}$

$$\lambda_k(t+1) = \lambda_k(t) + c(\beta(t+1) - \beta_k(t+1))$$

and for all $k \in \mathcal{N}, j \in \mathcal{N}_k$

$$\Lambda_j^k(t+1) = \Lambda_j^k(t) + c\left(P_j(t+1) - P_j^k(t+1)\right),$$

Here $P_j^k(t+1)$ is the component of the feasible tuple $Y_k(t+1)$ computed at the previous step.

multipliers $\lambda^k(0) = 0, \Lambda_j^k(0) = 0$. With these initial data, the tuple Y_k and the Lagrangian multipliers λ_k, Λ_j^k are iterated according to Algorithm 1. The algorithm involves a step size $c > 0$, which can be assigned as a constant or a time-varying function.

As mentioned before, the key issue in Algorithm 1 is to ensure that Step 2 is attainable since Ω_k is not compact. This issue is addressed in the following Lemma.

Lemma 3.1 *Let Assumption 3.3 be satisfied. Step 2 in Algorithm 1 is attainable if and only if Ω_k is nonempty.*

Proof. If Ω_k is empty, Step 2 is clearly not attainable, since the optimization over $Y_k \in \Omega_k$ cannot be carried out.

Conversely, suppose that Ω_k is nonempty. Since the inequalities (3.20) and (3.28)-(3.29) are not strict, Ω_k is a closed set. We now show that the set Ω_k is bounded in the variables P_k^k, F^k, and for the remaining variables, we will establish a coercion argument to ensure that the minimization in Step 2 does not make these variables diverge.

Select $Y_k \in \Omega_k$. Using the Schur complement, it follows from (3.28) that

$$-\rho_k P_k^k + F_k^\top (P_k^k)^{-1} F_k \preceq 0, \tag{3.32}$$

and it follows from the left-upper block of (3.24) that

$$Q_k + W_k + \frac{\psi_k^2}{\beta_k} C_k^\top C_k + \frac{1}{\beta_k} P_k^k B^w B^{w\top} P_k^k \preceq 0.$$

With (3.27) and adding (3.32), we have

$$\begin{aligned} P_k^k A + A^\top P_k^k - 2\psi_k C_k^\top C_k + \frac{\psi_k^2}{\beta^k} C_k^\top C_k - p_k F_k - p_k F_k^\top + \alpha_k P_k^k + W_k \\ + p_k \pi_k P_k^k + \frac{1}{\beta^k} P_k^k B^w B^{w\top} P_k^k - \tau_k \rho_k P_k^k + \tau_k F_k^\top (P_k^k)^{-1} F_k \preceq 0, \end{aligned}$$

where $\tau_k > 0$ is an arbitrary constant. After completion of squares, this leads to the inequality

$$\begin{aligned} P_k^k A + A^\top P_k^k - 2\psi_k C^{k\top} C^k + \frac{\psi_k^2}{\beta_k} C_k^\top C_k + (\alpha_k - \tau_k \rho_k) P_k^k + W_k \\ + p_k \pi_k P_k^k + \frac{1}{\beta_k} P_k^k B^w B^{w\top} P_k^k - \frac{p_k^2}{\tau_k} P_k^k + \left(F_k^\top - \frac{p_k}{\tau_k} P_k^k \right) \tau_k (P_k^k)^{-1} \left(F_k - \frac{p_k}{\tau_k} P_k^k \right) \preceq 0. \end{aligned}$$

Letting $\tilde{\tau}_k = \frac{1}{2}(\alpha_k - \tau_k \rho_k - \frac{p_k^2}{\tau_k})$ and dropping some positive semi-definite terms, allows us to conclude that

$$P_k^k (A + \tilde{\tau}_k I) + (A^\top + \tilde{\tau}_k I) P_k^k - 2\psi_k C^{k\top} C^k + \frac{1}{\beta^k} P_k^k E^k E^{k\top} P_k^k \preceq 0.$$

Multiplying the inequality from left and right with $X^k = (P_k^k)^{-1}$ results in

$$(A + \tilde{\tau}_k I) X^k + X^k (A^\top + \tilde{\tau}_k I) - 2\psi_k X^k C^{k\top} C^k X^k + \frac{1}{\beta^k} E^k E^{k\top} \preceq 0. \tag{3.33}$$

Now consider the algebraic Riccati equation obtained from (3.33) by replacing the inequality sign with equality,

$$(A + \tilde{\tau}_k I) X_0^k + X_0^k (A^\top + \tilde{\tau}_k I) - 2\psi_k X_0^k C^{k\top} C^k X_0^k + \frac{1}{\beta^k} E^k E^{k\top} = 0. \tag{3.34}$$

From Assumption 3.3, it follows that $(A + \tilde{\tau}_k I, B^w)$ is controllable, therefore $(A^\top + \tilde{\tau}_k I, B^{w\top})$ is observable. Moreover, $\tau_k > 0$ can be defined such that $(A^\top + \tilde{\tau}_k I, C_k^\top)$

is stabilizable. This can even be done when (A, C_k) is not detectable by choosing τ_k large enough. Then, we conclude that with this suitable choice of τ_k, (3.34) has a unique, positive definite, stabilizing solution X_0^k. Furthermore, for all solutions of the inequality (3.33), it holds that $X^k \geq X_0^k$ and subsequently, $P_k^k \preceq (X_0^k)^{-1}$. The solution variable P_k^k is therefore bounded. Moreover, from (3.28) and boundedness of P_k^k, we can conclude boundedness of F^k.

In constrast, the variables P_j^k for $j \in \mathcal{N}_k$ and β_k are not restricted to a bounded set by the LMIs under consideration. However, note that the cost function in Step 2 is quadratic in β_k and P_j^k for $j \in \mathcal{N}_k$. Thus, due to the boundedness of F_k, P_k^k, we conclude that the sub-level sets of the optimization objective function in Step 2

$$\left\{ Y_k \in \Omega_k | \ \beta_k - \lambda_k \beta_k + \frac{c}{2}|\beta - \beta_k|^2 + \sum_{j \in \overline{\mathcal{N}}_k} \left(-tr\left(\Lambda_j^{k\top} P_j^k\right) + \frac{c}{2}\|P^j - P_j^k\|^2 \right) < \overline{c} \right\} \tag{3.35}$$

are bounded for all $\overline{c} > 0$. We can conclude that we can equivalently search for the minimum of the cost function over a non-empty sub-level set (3.35) instead of Ω_k. Therefore, we can conclude that Step 2 in Algorithm 1 is always attainable. ■

In Algorithm 1, Steps 2 and 3 can be carried out in parallel; these steps do not require exchange of information between the agents. Step 1 needs a two step transmission process to calculate $P_j(t+1)$, where all agents j collect the respective matrices P_j^k for $k \in \mathcal{N}_j$, calculate $P_j(t+1)$, and transmit it to the neighbouring agents. To calculate $\beta(t+1)$, an average consensus algorithm (Olfati-Saber et al., 2007) can be applied when Assumption 3.2 is satisfied. For practical purposes, discrete time algorithms such as (Zhu & Martínez, 2010) or algorithms which converge in finite-time (Sundaram & Hadjicostis, 2007; L. Wang & Xiao, 2010; Chen et al., 2011) may be useful here.

If β is prescribed as a fixed performance parameter as discussed in Remark 3.4, then the calculation of $\tilde{\beta}$ is not required in Step 1 and all β-related terms in Step 2 can also be removed.

Remark 3.5 *Assumption 3.2 was made in order to enable step 1. In cases, where Assumption 3.2 is not satisfied, an alternative way to enable step 1 must be found depending on the underlying communication topology. Some examples are given below:*

- *Cycle-free graphs: This is the "trivial" case, as no iterations are necessary. Every observer only needs to wait for its neighbours $P_j^j, j \in \mathcal{N}_k$ and subsequently calculate P_k^k.*

- *Strongly-connected balanced graphs: In this case, if every observer k stores its representation P_j^k for all $j \in \mathcal{N}$, then every P_j can be calculated by average consensus.*

3.2.3 Simulation example

To illustrate the iterative algorithm, we take the example from (Ugrinovskii, 2011), which is a system of the form (2.1), with

$$A = \begin{bmatrix} 0.3775 & 0 & 0 & 0 & 0 & 0 \\ 0.2959 & 0.3510 & 0 & 0 & 0 & 0 \\ 1.4751 & 0.6232 & 1.0078 & 0 & 0 & 0 \\ 0.2340 & 0 & 0 & 0.5596 & 0 & 0 \\ 0 & 0 & 0 & 0.4437 & 1.1878 & 0.0215 \\ 0 & 0 & 0 & 0 & 2.2023 & 1.0039 \end{bmatrix},$$
$$B^w = \begin{bmatrix} 0.1\, I_6 & 0 \end{bmatrix}.$$

Six observers measure two coordinates each and none of the observers is able to estimate the complete state vector by itself. The communication topology is assumed to be the directed circulant graph as show in Figure 3.1 and we use Algorithm 1 to calculate the filter gains.

Since we are dealing with a directed but balanced graph, we apply the method described in Remark 3.5 and use complete local representations of all variables X_j at every observer k. The algorithm is run with fixed performance parameter $\beta^{const} = 1$ as discussed in Remark 3.4. We evaluate the matrix convergence by calculating the average value $X_j^{ave} = \frac{1}{N}\sum_{k\in\mathcal{N}} X_j^k$ and subsequently Error $= \sum_{j=1}^{N}\sum_{k=1}^{N} \|X_j^k - X_j^{ave}\|^2$. The resulting plot of the error evolution is shown in Figure 3.5.

3.3 Self-adapting solution with Ricatti-differential equations

The LMI-based solutions of the distributed estimation problem that we discussed so far all use some pre-computation to determine the observer gain matrices in (2.6) for estimating linear systems or its extended version (2.31) for systems with additive nonlinearity. The great benefit of this approach is that after the pre-computation, the observers have a simplistic form and do not require much computation. The first part of this chapter was

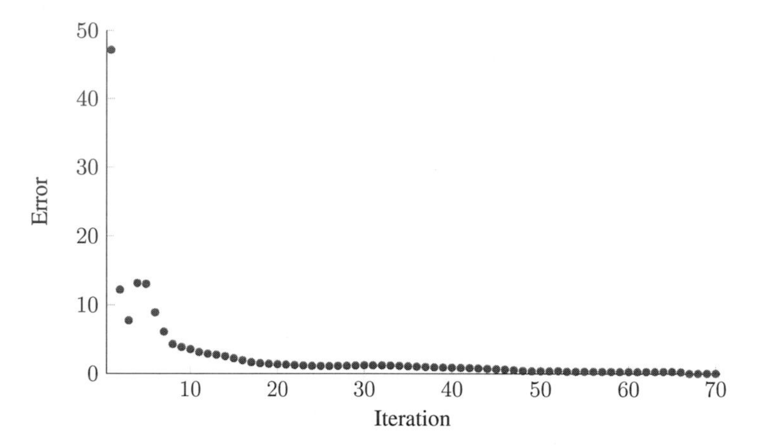

Figure 3.5: Evolution of the error during iteration for fixed β.

dedicated at enabling distributed design for the observer gain matrices in order to improve applicability.

In contrast to the methods presented so far, *Distributed Kalman Filtering* takes a very different approach and proposes that the observer gain matrices are dynamically adapted, without spending much effort on pre-computation. This method was adapted from classical filtering theory (Kalman, 1960) and important results where published in (Olfati-Saber, 2005, 2006, 2007; Carli et al., 2008; Kamal et al., 2013), to name a few. In the following, we discuss the idea of Distributed Kalman Filtering and discuss our recent results for Distributed Kalman Filtering in continuous-time systems presented in (Wu, Elser et al., 2016)[1] and (J. Kim, Shim & Wu, 2016).

As this short literature review above shows, many results are derived for discrete-time systems in the form of

$$x(t+1) = Ax(t) + B^{w}w(t)$$

with N observers, which each measure some output

$$y_k(t) = C_k x(t) + \eta_k(t), \quad k \in \mathcal{N}.$$

Here, $t \in \mathbb{N}$ and $w(t), \eta_k(t), k \in \mathcal{N}$ are uncorrelated zero-mean white Gaussian noise. In this case, the individual observers can be written in form of a prediction-correction (or

[1] This paper mainly resulted from the collaboration with Anja Elser, who worked on this topic for her Master thesis.

correction-prediction) procedure, as it is common for Kalman-Filtering (Kalman, 1960). In particular, the correction step takes the interaction between the observers into account, where the interaction both affects the estimates $\hat{x}_k$ and the covariance matrices P_k. For instance, in (Olfati-Saber, 2007), the distributed observers are proposed as

Covariance update **Estimate update**

$$
\begin{aligned}
S_k &= \sum_{j \in \mathcal{N}_k \cup i} C_j^\top R_j^{-1} C_j & z_k &= \sum_{j \in \mathcal{N}_k \cup i} C_j^\top R_j^{-1} z_j \\
M_k &= (P_k^{-1} + S_k)^{-1} & \overline{x}_k &= \hat{x}_k + M_k(z_k - S_k \overline{x}_k) + k M_k \sum_{j \in \mathcal{N}_k} (\hat{x}_j - \hat{x}_k) \\
P_k &\leftarrow A M_k A^\top + B^w Q B^{w\top} & \hat{x}_k &\leftarrow A \overline{x}_k
\end{aligned}
\tag{3.36}
$$

Some extensions to this approach are presented for instance in (Kamal et al., 2013), where particularly some observability assumptions are relaxed.

In the case of estimating continuous-time LTI systems

$$
\begin{aligned}
\dot{x} &= Ax + B^w w \\
y_k &= C_k x + \eta_k, \quad k \in \mathcal{N}
\end{aligned}
\tag{3.37}
$$

our goal is to achieve Property 1 of Problem 1, under Assumptions 3.2, 3.3, and

Assumption 3.4 *The pair* $(A, [C_k]_{k \in \mathcal{N}})$ *is observable.*

Here, we assume observability for technical reasons. If the pair $(A, [C]_{k \in \mathcal{N}})$ is detectable, but not observable, the unobservable component can be removed by Kalman decomposition (Kalman, 1962), in order to apply the following methods.

For continuous-time systems, it is well known that instead of the prediction-correction procedure, the observers incorporate the solution of an Riccati Differential Equation (RDE) (Kalman & Bucy, 1961) of the form

$$
\dot{P} = AP + PA^\top + B^w Q B^{w\top} - PC^\top R^{-1} CP, \tag{3.38}
$$

where $P(t) \in \mathbb{R}^{n \times n}$ is positive definite for all $t > 0$. The solution of the RDE can be subsequently implemented to determine the correction gain matrix $L(t)$. Centralized observers that follow this scheme are referred to as *Kalman-Bucy Filter*. Hence, we will call this type of distributed estimation *Distributed Kalman-Bucy Filtering*.

In both (2.6) and (3.36), couplings between the estimates were applied. Therefore, it is safe to assume that they will again be used for the estimation of (3.37). However, the essential questions is how the interaction between the distributed observers will affect the individual RDEs. In the following, we will discuss a number of options.

3.3.1 Simple coupling of RDEs

First, we discuss the complete absence of couplings between the RDEs. In (Olfati-Saber, 2007) the distributed observers are porposed as

$$\dot{\hat{x}}_k = A\hat{x}_k + L_k(t)(y_k - C_k\hat{x}_k) + \gamma \sum_{j\in N_k} (\hat{x}_j - \hat{x}_k) \tag{3.39}$$

$$\begin{aligned} L_k(t) &= P_k(t)C_k^\top R_k^{-1} \\ \dot{P}_k &= AP_k + P_kA^\top + B^wQB^{w\top} - P_kC_k^\top R_k^{-1}C_kP_k, \end{aligned} \tag{3.40}$$

with $\gamma > 0$. These distributed observers in fact resemble the centralized observer as proposed in (Kalman & Bucy, 1961), with the additional diffusive term $\gamma \sum_{j\in N_k}(\hat{x}_j - \hat{x}_k)$ in (3.39). Now suppose the pair (A, C_k) is undetectable for some $k \in \mathcal{N}$. Then, in (Callier & Willems, 1981; Callier & Winkin, 1995), the authors present conditions under which the RDE converges. However, there are also cases, where P_k in (3.40) diverges exponentially, which in turn causes divergence of the observer gains matrices L_k. The simplest example for the latter case is a decoupled system with

$$A = \begin{bmatrix} A_1 & 0 \\ 0 & A_2 \end{bmatrix} \quad C_k = \begin{bmatrix} \widetilde{C} & 0 \end{bmatrix}, \tag{3.41}$$

where A_2 is unstable. This issue limits the applicability of (3.39), which illustrates why couplings between the RDEs should be introduced.

The most intuitive way of adding couplings to (3.40) is by implementing diffusive terms in the RDEs in the form of $k \sum_{j\in N_k}(P_j - P_k)$. Thus, the distributed observers read

$$\dot{\hat{x}}_k = A\hat{x}_k + NL_k(y_k - C_k\hat{x}_k) + \gamma \sum_{j\in N_k} (\hat{x}_j - \hat{x}_k) \tag{3.42}$$

$$\begin{aligned} L_k &= P_kC_k^\top R_k^{-1} \\ \dot{P}_k &= AP_k + P_kA^\top + B^wQB^{w\top} - NP_kC_k^\top R_k^{-1}C_kP_k + k\sum_{j\in N_k}(P_j - P_k) \end{aligned} \tag{3.43}$$

with coupling strengths $\gamma, k > 0$. This approach is presented in (J. Kim, Shim & Wu, 2016), and it can be shown that for $\gamma, k \to \infty$ and $t \to \infty$, the solutions of the coupled RDEs (3.43) approach the solution of the centralized RDE (3.38) for $C = [C_k]_{k\in\mathcal{N}}$. Thus, the optimality properties of the centralized Kalman-Bucy-Filter (Kalman & Bucy, 1961) are recovered for $\gamma, k \to \infty$.

However, since the coupling strengths cannot be chosen infinitely large for computational reasons, it becomes important to give a lower bound on γ and k, such that estimates are well-behaved. Calculation of these lower bounds is challenging, as shown in (J. Kim, Shim & Wu, 2016). In particular, the sufficient condition for the coupling strengths takes the explicit solutions $P_k(t)$ of (3.43) for all $k \in \mathcal{N}$ into account, which makes it an implicit condition and therefore needs to be approximated by a conservative bound.

3.3.2 Information weighted coupling of RDEs

In the following we present an alternative approach, which does not include the calculation of coupling strengths as seen in (3.42), (3.43).

Again, we consider the LTI system (3.37) with the measured outputs y_k. The distributed observers are proposed as

$$\dot{\hat{x}}_k = A\hat{x}_k + L_k(y_k - C_k\hat{x}_k) + P_k \sum_{j\in N_k} P_j^{-1}(\hat{x}_j - \hat{x}_k) \tag{3.44}$$

$$L_k = P_k C_k^\top R_k^{-1}$$

$$\dot{P}_k = AP_k + P_k A^\top + B^w Q B^{w\top} - P_k C_k^\top R_k^{-1} C_k P_k - P_k \sum_{j\in N_k} (P_j^{-1} - P_k^{-1})P_k, \tag{3.45}$$

where every P_k is initialized with $P_k(0) = P_{k,0} \succ 0$. In contrast to (3.39) and (3.42), the diffusive state coupling $\hat{x}_j - \hat{x}_k$ is weighted by $P_k P_j^{-1}$. This means, for instance, that $(\hat{x}_j - \hat{x}_k)$ is weighted lower when $P_k < P_j$, i.e. if the estimate $\hat{x}_k$ is more confident than the estimate $\hat{x}_j$. A similar intuition is applicable to (3.45). This kind of additional weighting was used for discrete-time distributed Kalman Filtering in (Kamal et al., 2013). In particular, the novel coupling term in (3.45) ensures that P_k do not diverge even if the pairs (A, C_k) are not detectable.

In order to analyze the behavior of (3.45), it is helpful to consider the inverse RDE in terms of $W_k(t) := P_k^{-1}(t)$, given by

$$\begin{aligned}\dot{W}_k &= - W_k \dot{P}_k W_k \\ &= - W_k A - A^\top W_k - W_k B^w Q B^{w\top} W_k + C_k^\top R_k^{-1} C_k + \sum_{j\in N_k} (W_j - W_k).\end{aligned} \tag{3.46}$$

Note that (3.46) is diffusively coupled, same as (3.43), but where (3.43) relies on the fact that for increasing coupling strength $k, \gamma \to \infty$ and $t \to \infty$, all $P_k(t)$ converge towards the centralized solution, (3.46) does not rely on an argument like that. Instead, distinct steady states of W_k and P_k, respectively, are acceptable.

The convergence of coupled RDE such as (3.46) is a complex, non-standard problem, for which results in the literature are very limited. In (Jodar & Hervas, 1989), some results were presented that show guarantees for time intervals, where the solution to the coupled RDEs exists and how to approximate the solution. In (Abou-Kandil et al., 1993, 1994; Freiling et al., 1996), the two player-model of coupled RDE was analysed. A complete solution to analysing the evolution of (3.46) is yet to be found, but in the following, we will give some results on the behaviour of (3.46) and argue, why this is a suitable approach.

First, we note that if $W_k, k \in \mathcal{N}$ are initialized as positive definite matrices, then we have $W_k(t) \succ 0$ at every time instance $t > 0$. This results can e.g. be deducted from (Dieci & Eirola, 1994). Now, we consider the tuple of all $W_k, k \in \mathcal{N}$ and for the steady states of (3.46), if there are any, we can show the following lemma:

Lemma 3.2 *Given the matrix differential equations* (3.46) *and suppose that the tuple* $(W_1^*, ..., W_N^*)$ *represents a steady state, i.e.* $\dot{W}_k = 0$ *for all* $k \in \mathcal{N}$ *if* $W_k = W_k^*, k \in \mathcal{N}$. *Then, it holds that* $W_k^* \succ 0$ *for all* $k \in \mathcal{N}$.

Proof. *For the proof, please see (Wu, Elser et al., 2016).* ■

From Lemma 3.2 we know that there is no steady state for the matrices W_k, where even one of the matrices W_k becomes singular. Together with the fact that $W_k(t) \succ 0$ for all t, this justifies the following assumption:

Assumption 3.5 *For all* $k \in \mathcal{N}$, *there exists a positive definite lower bound matrix* $S_k \succ 0$ *in the sense that* $P_k^{-1}(t) = W_k(t) \succ S_k \succ 0$ *for all* $t \geq 0$.

This assumption is important as it enables the usage of an invariance-type argument in order to proof convergence of the estimates, as shown in the following Theorem.

Theorem 3.2 *Consider the system (3.37) and the observers* (3.44), (3.45), *with communication topology represented by the connected graph* $\mathcal{G}$. *Let Assumptions 3.2, 3.3, 3.4, and 3.5 hold true. Then Property 1 of Problem 1 holds.*

Proof. *For the proof, please see (Wu, Elser et al., 2016).* ■

Next, we show a simulation example, which illustrates Assumption 3.5 and the convergence of the estimates.

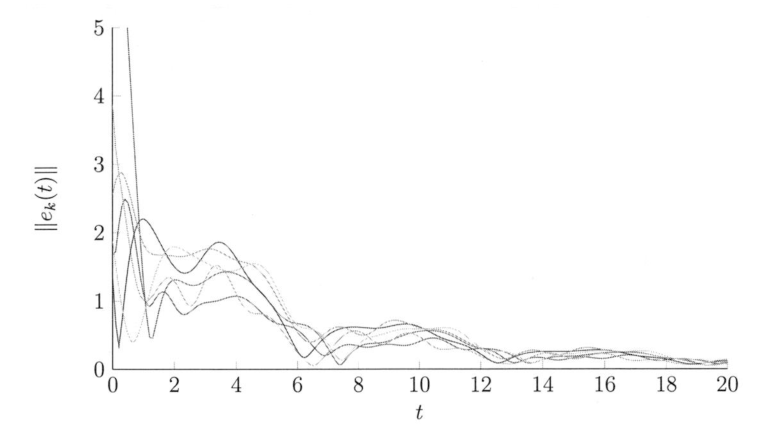

Figure 3.6: Plots of the estimation error of all observers.

3.3.3 Simulation example

We consider the six-dimensional unstable oscillator given by

$$\dot{x} = \begin{bmatrix} .1 & 1 & 0 & 1 & 0 & 1 \\ -1 & .1 & 1 & 0 & 1 & 0 \\ 0 & -1 & .1 & 1 & 0 & 1 \\ -1 & 0 & -1 & .1 & 1 & 0 \\ 0 & -1 & 0 & -1 & .1 & 1 \\ -1 & 0 & -1 & 0 & -1 & .1 \end{bmatrix} x + \begin{bmatrix} 0 \\ 1 \\ 0 \\ 1 \\ 0 \\ 1 \end{bmatrix} w.$$

The observer communication topology and the respective output matrices C_k are chosen identically to Section 2.2.6, shown in Figure 2.3. Again, for all $k \in \mathcal{N}$, (A, C_k) is undetectable.

The result after applying the distributed observers (3.44), (3.45) are shown in Figure 3.6 and 3.7. In particular, Figure 3.7 illustrates that in our example, the matrices P_k reach steady state, which also ensures that Assumption 3.5 is satisfied. Moreover, it should be noted that P_k are not required to reach consensus, which can also be seen in Figure 3.7.

3.4 Summary and Discussion

In this section, we have further enhanced the results from Chapter 2 by developing methods that allow to reduce the required communication bandwidth and apply distributed

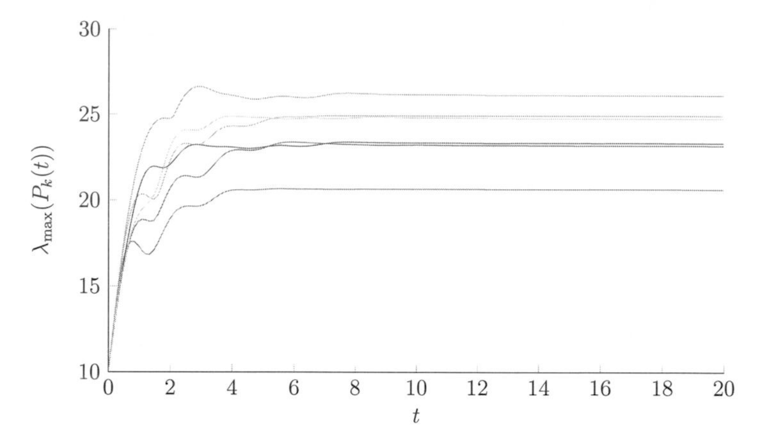

Figure 3.7: Maximum eigenvalues of P_k following the RDE.

design methods for computing the observer gain matrices in (2.6). Both parts are essential for the applicability of distributed estimation and have not been solved so far to the best of our knowledge.

Reducing transmission was achieved by introducing event-triggered broadcasting of observer states in the distributed estimation scheme from Chapter 2. Triggering the events is done by a rule that can be monitored by the observers individually. Therefore, the perfect, continuous transmission of observer states, as it was assumed in Chapter 2, is not needed anymore. Moreover, we thoroughly discussed the inter-event times for the case with exogenous disturbances and measurement noise. Under the mild assumption of bounded disturbances, we were able to exclude Zeno behaviour and moreover were able to present a direct relationship between various system parameters and a lower bound on the persistent inter-event times. It should be noted that this improvement with respect to the required bandwidth comes at the expense of higher computational load at every individual observer. Therefore, it depends on the application, whether event-triggered distributed estimation is a preferable solution or not. As an alternative, periodic communication can be considered, which corresponds to our event-triggered scheme with fixed inter-event times. Here, the challenge is to find a period length that is small enough to guarantee stability, but large enough to be reasonable for practical applications. Some loosely related results in this direction have been published e.g. in (Dörfler et al., 2013; Açikmese et al., 2014). Moreover, our results on event-triggered communication can be extended towards self-triggering, where the observers do not need to constantly monitor a trigger rule, but instead predict the next triggering time instance. Some references on self-triggered control, among many others, are (X. Wang & Lemmon, 2009; Anta & Tabuada, 2010).

For achieving distributed design of the observer gain matrices, two fundamentally different ways were discussed: The first way directly follows the LMI-based design conditions discussed in Chapter 2 and builds upon technique from distributed optimization, in order to decouple the LMI conditions and enable parallel computation. As a matter of fact, in (Ugrinovskii, 2011) a gradient-descent-type algorithm was proposed that can also be used to calculate the filter gains in a distributed manner. Although the proposed gradient type algorithm demonstrated a possibility of computing the observer gain matrices in principle, a practical application of that algorithm is hindered due to slow convergence observed even in low dimensional examples. In this respect the procedure presented in Section 3.2 turn out to be superior in simulations and more importantly, applies to well-known methods from distributed optimization, which allows for further investigation on convergence speed, improved iterations, or continuous-time algorithms.

The second way involves the implementation of coupled Riccati Differential Equations. Compared to the first way, this method causes higher computational load as well as higher communication load, but do not require any offline-solution of LMI-condition. Instead, the observer gain matrices are adapted on-the-fly. We discussed a number of different ways to implement the couplings between the RDEs and their respective advantages. One thing, which was not discussed in Section 3.3.2 was the performance with respect to the disturbances or optimality of the estimates, respectively. Unlike the averaging approach from Section 3.3.1, optimality of the estimates can not directly be established, and is therefore open for future research.

In many applications, where changes in the communication topology or to the system itself may happen occasionally, the methods presented in Sections 3.2 and 3.3 both give important ways to adapt the distributed observers. For instance, consider the case where one observer with some C_k fails and is replaced with another observer with $C_j \neq C_k$. Then both the procedures in Section 3.2 and Section 3.3 allow the observers to reconfigure using local communication and computation only. Note that however, a short disruption to the estimation errors is sure to happen. If changes to the observer structure of the estimated system happen frequently, then these kind of switches need to be taken into account explicitly for the observer design procedure. In (Ugrinovskii, 2013), some extensions towards randomly switching networks are presented.

Chapter 4

Distributed $\mathcal{H}_\infty$ State Estimation-based Multi-Agent Coordination

In the previous chapters we have significantly extended the class of systems that can be considered for distributed estimation and developed improvements to communication and computation. Furthermore, in the introduction of this thesis, we have discussed that distributed estimation algorithms enable new ways to apply distributed control algorithms. This chapter is dedicated at applying these techniques to the control of multi-agent systems (MAS), and more specifically to solve synchronization and output regulation tasks, which are two of the most important tasks within the field of multi-agent coordination.

As also discussed in the introduction of this thesis, research on synchronization problems has become one of the major fields in the research on control of autonomous agents. In (Wieland, Sepulchre & Allgöwer, 2011), the synchronization problem for hetergeneous MAS was solved by locally enforcing a common trajectory upon the agents, which however relies on the assumption that every agent on its own is able to estimate its own state. However, measurements may depend on the state of more than one agent, for example, if distance between the agents are measured. In this case of *relative measurements*, it was shown in (Grip et al., 2012b, 2012a, 2015) that synchronization can be achieved under the restriction to leader-follower graphs. In this chapter, we present solutions to the synchronization problem without imposing this assumption on the topology. Instead, we present two approaches, where the first is based on expanding the agents' spectrum and the second is based on applying distributed estimation technique as discussed above. However, to pursue the latter approach, the distributed estimation scheme introduced in Chapter 2 needs to be adapted in order to fit the application to large-scale systems: So far we were dealing with single, unstructured systems, and specifically, all observers were assigned to estimate the system's full state. When dealing with large-scale systems such as multi-agent systems (MAS), then the previous solutions are not suitable for estimation because of one main reason: In large-scale systems, algorithms for estimation and control are supposed to be scalable in the sense that computational demand is distributed over a number of agents. In particular, for growing size of the system, the number of agents may grow, but complexity of computation at every agent is not supposed to grow. The algorithms for distributed estimation in the previous chapter do not satisfy this requirement as the dimension of each design LMI and each observer grows with the number of agents N. Hence, in this chapter, we will specifically consider large-scale systems, consisting of autonomous agents or interconnected agents, and alter the distributed estimation setup from the previous chapter in order to preserve scalability when estimating systems of this class.

This chapter is structured as follows: In Section 4.1, the Internal Model Principle for synchronization is revisited and a geometric solution to synchronization is presented that can be applied under certain assumptions on the agents. This part is based on (Wu & Allgöwer, 2012). Then, in Section 4.2, the observer-based approach is presented where cooperating observers provide the state estimates needed for solving the synchronization problem and thereby achieving performance guarantees with respect to exogeneous disturbances and measurement noise. This part is based on (Wu et al., 2014, 2017). Moreover, these results can be extended towards physically interconnected agents, which is shown in Section 4.3. Further, in Section 4.3, we present two more extensions of the observer-based scheme, where our focus lies on relaxing the requirement of model information availability and considering exosystems generating disturbances or references. This part is based on (Wu, Ugrinovskii & Allgöwer, 2016; Wu & Allgöwer, 2016b).

4.1 The Synchronization Problem

We consider a group of N LTI agents described by the differential equation

$$\dot{x}_k = A_k x_k + B_k u_k \tag{4.1}$$

where for all $k \in \mathcal{N}$, $x_k(t) \in \mathbb{R}^{n_k}$ is the state of the agent and $u_k(t) \in \mathbb{R}^{m_k}$ is the respective control input. As in Chapter 2, the communication topology between the agents is defined by the directed graph $\mathcal{G} = (\mathcal{V}, \mathcal{E})$, i.e. agent k may receive information from agent j if $(v_j, v_k) \in \mathcal{E}$. The outputs to be synchronized are defined as

$$z_k = \widetilde{C}_k x_k, \tag{4.2}$$

where $z_k(t) \in \mathbb{R}^{\tilde{r}}$ for all $k \in \mathcal{N}$. Thus, we have the definition of synchronization:

Definition 4.1 *Let there be a multi-agent system consisting of N agents* (4.1) *with the outputs* (4.2). *The agents are said to* synchronize, *if*

$$\lim_{t \to \infty} z_k - z_j = 0 \quad \textit{for all } j, k \in \mathcal{N}. \tag{4.3}$$

Moreover, as discussed in the introduction to this chapter, we consider the case where relative output sensing is applied, i.e. the measurement outputs are

$$y_k = \sum_{j \in \mathcal{N}_k} (z_j - z_k). \tag{4.4}$$

With (4.4), we implicitly made the following assumption on the communication topology.

Assumption 4.1 *If the measurement output y_k of agent k is affected by $x_j, j \neq k$, then agent k may receive information from agent k, i.e. $(v_j, v_k) \in \mathcal{E}$.*

Concerning the individual agent (4.1), we make the following basic assumption

Assumption 4.2 *The pairs (A_k, B_k) are stabilizable and the pairs $(A_k, \widetilde{C}_k)$ are detectable.*

With the definition of the system class under consideration we can now propose the closed-loop synchronization problem that is addressed in this chapter.

Problem 4 (Closed loop output synchronization) *For every agent k, determine a dynamic controller, which receives the measurement y_k and data communicated from the neighbouring agents $j \in \mathcal{N}_k$, such that synchronization is achieved.*

In the remainder of this section, two fundamentally different approaches to address synchronization are presented and discussed.

4.1.1 Synchronization by model expansion

In the presence of non-identical agents (4.1) and the task to synchronize their outputs z_k, it is an essential question what common trajectory they can synchronize to, other than the trivial case of all z_k converging towards 0. This question was thoroughly answered by the *Internal Model Principle for Synchronization (IMP)* in a number of publications, where both linear agents and nonlinear agents where considered. Important references herein are (Wieland & Allgöwer, 2009; Wieland, Sepulchre & Allgöwer, 2011; H. Kim et al., 2011;

Lunze, 2011) and (Wieland et al., 2013). Essentially, the ability of heterogeneous agents to synchronize is closely related to the existence of a solution to the Regulator Equations

$$\begin{aligned} A_k\Pi_k + B_k\Gamma_k &= \Pi_k S \\ \widetilde{C}_k\Pi_k &= R, \end{aligned} \tag{4.5}$$

for all $k = 1, ..., N$, where

$$(S, R) \text{ is observable and } \sigma(S) \subset \overline{\mathbb{C}}^+. \tag{4.6}$$

The Regulator Equations are well known from the theory of Output Regulation (Francis & Wonham, 1975, 1976; Knobloch et al., 1993) and for the synchronization problem it means that there must exist a pair (S, R) such that "all individual systems are able to track the same virtual exosystem defined by the state feedback matrix S and the output matrix R" (Wieland, Sepulchre & Allgöwer, 2011). Since solvability of (4.5) with (S, R) being observable is a necessary condition for synchronization of the unperturbed systems (4.1), assuming the existence of a solution to (4.5) is not conservative.

Note that the solution (S, R) is far from being unique and in this section, we present an approach to tackle the synchronization problem by defining S as an appropriate *union* of the persistent components of A_k .

For this approach, we consider agents satisfying the following assumption:

Assumption 4.3 *The spectrums of all agents lie in the closed left-half complex plane $\sigma(A_k) \subset \overline{\mathbb{C}}^-$.*

This assumption is made because exponential divergence of the agents' states x_k is commonly undesirable and the method presented in this section aims at synchronization by expanding the models instead of pole placement. For the algebraic multiplicity of the eigenvalues on the imaginary axis, we denote as $\overline{\alpha}_A(\lambda_i)$ the algebraic multiplicity of an eigenvalue λ_i of the matrix A and introduce the notation

$$\eta_k = \sum_{\lambda \in (\sigma(A_k) \cap i\mathbb{R})} \overline{\alpha}_{A_k}(\lambda). \tag{4.7}$$

The goal now is to find suitable matrices $S \in \mathbb{R}^{\nu\times\nu}$, $R \in \mathbb{R}^{\tilde{r}\times\nu}$ such that

- S, R satisfy (4.6), and in particular $\sigma(S) \subset i\mathbb{R}$.
- For all $k \in \mathcal{N}$, there exists a matrix $\tilde{S}_k$ with $\sigma(\widetilde{S}_k) \subset i\mathbb{R}$ that is complementary to A_k in the sense that

$$(\sigma(A_k) \cup \sigma(\tilde{S}_k)) \cap i\mathbb{R} = \sigma(S). \tag{4.8}$$

Furthermore, for all $\lambda \in \sigma(S)$ we aim at

$$\overline{\alpha}_{A_k}(\lambda) + \overline{\alpha}_{\widetilde{S}_k}(\lambda) = \overline{\alpha}_S(\lambda). \tag{4.9}$$

Thus, we clearly have $\nu \geq \eta_k$ for all $k \in \mathcal{N}$. The idea behind this procedure is homogenization of the agents by expanding the agents' state feedback matrix A_k through feedforward control, such that every extended matrix

$$\widetilde{A}_k = \begin{bmatrix} A_k & B_k\widetilde{G}_k \\ 0 & \widetilde{S}_k \end{bmatrix} \tag{4.10}$$

with some $\widetilde{G}_k$ can be transformed to

$$T_k^{-1}\widetilde{A}_kT_k = \begin{bmatrix} S & \widetilde{P}_k \\ 0 & \widetilde{Q}_k \end{bmatrix} \tag{4.11}$$

by using some transformation matrix T_k and some residual matrices $\widetilde{Q}_k, \widetilde{P}_k$.

The geometric conditions for the feasibility of this extension are formalized in the following Lemma.

Lemma 4.1 *Suppose there exist a positive integer ν and matrices $S \in \mathbb{R}^{\nu\times\nu}$, $R \in \mathbb{R}^{\widetilde{r}\times\nu}$, $\Pi_k \in \mathbb{R}^{n_k\times\nu}$, $\Gamma_k \in \mathbb{R}^{m_k\times\nu}$ such that* (4.6) *and* (4.5) *are satisfied for all $k \in \mathcal{N}$. If for all $k \in \mathcal{N}$ there exist matrices $\widetilde{S}_k \in \mathbb{R}^{(\nu-\eta_k)\times(\nu-\eta_k)}$ and $\Theta_k \in \mathbb{R}^{(\nu-\eta_k)\times\nu}$ with*

$$\widetilde{S}_k\Theta_k = \Theta_kS \tag{4.12}$$

$$\mathit{rank}\begin{bmatrix} \Pi_k \\ \Theta_k \end{bmatrix} = \nu \tag{4.13}$$

$$\mathit{Ker}(\Gamma_k) \supset \mathit{Ker}(\Theta_k), \tag{4.14}$$

then (4.8) *and* (4.9) *hold and there exists $T_k, \widetilde{G}_k$ such that* (4.11) *holds. Moreover, $\widetilde{Q}_k$ is Hurwitz, and thus represents the agents' transient dynamics.*

Proof. Suppose all conditions of the Lemma are satisfied. Then, due to the kernel condition (4.14), we can choose $\widetilde{G}_k$ such that $\widetilde{G}_k\Theta_k = \Gamma_k$, where Θ_k is defined through (4.12). Further, with (4.12), the extended system matrix (4.10) satisfies the implicit form of the IMP

$$\begin{aligned} \widetilde{A}_k\Phi_k &= \Phi_kS \\ \begin{bmatrix} \widetilde{C}_k & 0 \end{bmatrix}\Phi_k &= R, \end{aligned} \tag{4.15}$$

for all $k = 1, ..., N$ with $\Phi_k = \begin{bmatrix}\Pi_k^\top & \Theta_k^\top\end{bmatrix}^\top$, where $\text{rank}(\Phi_k) = \nu$ according to (4.13). Now, let $\Phi_k^\perp \in \mathbb{R}^{(n_k+\nu-\eta_k)\times(n_k-\eta_k)}$ be a matrix, such that $\Phi_k^T \Phi_k^\perp = 0$ and $\text{rank}(\Phi_k^\perp) = n_k - \eta_k$, i.e. the subspace spanned by the columns of $\Phi_k^\perp$ is the orthogonal complement to the subspace spanned by the columns of Φ_k. Then, the transformation matrix $T_k = \begin{bmatrix}\Phi_k & \Phi_k^\perp\end{bmatrix}$ transforms $\widetilde{A}_k$ into an upper block-triangular matrix

$$T_k^{-1}\widetilde{A}_k T_k = \begin{bmatrix} S & \widetilde{P}_k \\ 0 & \widetilde{Q}_k \end{bmatrix}$$

with $\widetilde{Q}_k \in \mathbb{R}^{(n_k-\eta_k)\times(n_k-\eta_k)}$. Since a coordinate transformation preserves the eigenvalues of a matrix, it holds that

$$\sigma(S) \cup \sigma(\widetilde{Q}_k) = \sigma(T_k^{-1}\widetilde{A}_k T_k) = \sigma(\widetilde{A}_k) = \sigma(A_k) \cup \sigma(\widetilde{S}_k). \tag{4.16}$$

From Assumption 4.3 we know that $\sigma(A_k) \subset \overline{\mathbb{C}}^-$. Therefore, as $S \in \mathbb{R}^{\nu\times\nu}$, $\widetilde{S}_k \in \mathbb{R}^{(\nu-\eta_k)\times(\nu-\eta_k)}$ and the algebraic multiplicity of $\sigma(A_k) \cap j\mathbb{R}$ is η_k, $\sigma(\widetilde{Q}_k) = \sigma(A_k)\backslash j\mathbb{R}$, i.e. $\widetilde{Q}_k$ is a Hurwitz matrix. ■

If the conditions from Lemma 4.1 are met, than a control scheme for achieving Problem 4, based on the relative outputs (4.4) is proposed as follows: For all $k \in \mathcal{N}$, let there be dynamic controllers

$$\begin{aligned}
\dot{\epsilon}_k &= \left(\widetilde{A}_k + \begin{bmatrix} B_k\widetilde{H}_k \\ \widetilde{F}_k \end{bmatrix} \widetilde{K}_k\right)\epsilon_k + \Phi_k L(\xi_k - y_k) \\
\dot{\zeta}_k &= \widetilde{S}_k\zeta_k + \widetilde{F}_k\widetilde{K}_k\epsilon_k \\
u_k &= \widetilde{G}_k\zeta_k + \widetilde{H}_k\widetilde{K}_k\epsilon_k \\
\xi_k &= \sum_{j\in\mathcal{N}_k} ([\widetilde{C}_j \; 0]\epsilon_j - [\widetilde{C}_k \; 0]\epsilon_k)
\end{aligned} \tag{4.17}$$

with the matrices $\widetilde{S}_k, \widetilde{G}_k, \Phi_k$ given by Lemma 4.1. Moreover, $\widetilde{F}_k, \widetilde{H}_k, \widetilde{K}_k, L$ are matrices of appropriate size, which need to be designed. The following Theorem presents the design procedure.

Theorem 4.1 *Let a group of N agents be given as* (4.1) *with synchronization outputs* (4.2) *and relative output sensing* (4.4), *and let the communication topology be defined by a connected graph $\mathcal{G} = (\mathcal{V}, \mathcal{E})$. Suppose Assumptions 4.1 - 4.3 hold and all conditions of Lemma 4.1 are satisfied.*

Then, the local controllers (4.17) *solve Problem 4 if the matrix*

$$\widetilde{A}_k + \begin{bmatrix} B_k\widetilde{H}_k \\ \widetilde{F}_k \end{bmatrix} \widetilde{K}_k \text{ is Hurwitz} \tag{4.18}$$

and $L = -\delta P R^T$, where P is the solution of the algebraic Riccati Equation

$$SP + PS^T - \delta^2 P R^T R P + I_n = 0, \tag{4.19}$$

where $2 \cdot Re(\lambda_2(\mathcal{L}_\mathcal{G})) > \delta > 0$.

Note that given stabilizability of (A_k, B_k) from Assumption 4.2, we can choose $\widetilde{F}_k$, and $\widetilde{H}_k$ as matrices of appropriate size, such that $\left(\widetilde{A}_k, \begin{bmatrix} B_k\widetilde{H}_k \\ \widetilde{F}_k \end{bmatrix}\right)$ is stabilizable. Thus, (4.18) can be made Hurwitz easily.

Before proving Theorem 4.1 we introduce a lemma that deals with the synchronization of a network of N identical agents. Consider the agents

$$\dot{x}_k = Sx_k + u_k, \quad k = 1, ..., N \tag{4.20}$$

where $x_k(t) \in \mathbb{R}^n$ is the state and $u_k(t) \in \mathbb{R}^n$ is the input. Let $S \in \mathbb{R}^{n\times n}$ and $R \in \mathbb{R}^{\tilde{r}\times n}$, where (4.6) holds. Let the relative output be given as

$$\xi_k = \sum_{j\in\mathcal{N}_k} (Rx_j - Rx_k), \tag{4.21}$$

then we have the following result on the design of synchronizing controllers.

Lemma 4.2 *Let N identical agents be modeled by (4.20), (4.21) for $k = 1, ..., N$ and let the communication topology be defined by a connected graph $\mathcal{G} = (\mathcal{V}, \mathcal{E})$. If P is the solution to the algebraic Riccati equation* (4.19)*, where $2 \cdot Re(\lambda_2(\mathcal{L}_\mathcal{G})) > \delta > 0$, then the local controller*

$$u_k = -\delta P R^T \xi_k \tag{4.22}$$

leads to synchronization of the state vectors x_k, in the sense of $(x_j - x_k) \to 0$ for all j, k as $t \to \infty$.

Proof. Since $\mathcal{G}$ is connected, $\mathrm{Re}(\lambda_2(\mathcal{L}_\mathcal{G})) > 0$, which ensures that δ can be chosen properly. The proof can be directly derived from (Wieland, Kim & Allgöwer, 2011). ∎

Proof. (Theorem 4.1) Suppose that all conditions of Theorem 4.1 are satisfied and define the matrices

$$\widetilde{B}_k = \begin{bmatrix} B_k\widetilde{H}_k \\ \widetilde{F}_k \end{bmatrix}. \tag{4.23}$$

We denote the vector $\tilde{x}_k = \begin{bmatrix} x_k^T & \zeta_k^T \end{bmatrix}^T$ with $\zeta_k \in \mathbb{R}^{\nu-\eta_k}$, and apply (4.17) Thus, the closed-loop system with agent state variable $\tilde{x}_k$ and controller state variable ϵ_k can be written as

$$\begin{aligned}
\dot{\tilde{x}}_k &= \widetilde{A}_k \tilde{x}_k + \widetilde{B}_k \tilde{u}_k \\
\dot{\epsilon}_k &= (\widetilde{A}_k + \widetilde{B}_k \widetilde{K}_k)\epsilon_k + \Phi_k L(\xi_k - y_k) \\
\xi_k &= \sum_{j \in \mathcal{N}_k} ([\widetilde{C}_j \; 0]\epsilon_j - [\widetilde{C}_k \; 0]\epsilon_k) \\
\tilde{u}_k &= \widetilde{K}_k \epsilon_k .
\end{aligned} \tag{4.24}$$

Next, consider the coordinate transformation

$$s_k = T_k^{-1}(\tilde{x}_k - \epsilon_k). \tag{4.25}$$

The solution of s_k follows

$$\begin{aligned}
\dot{s}_k =& T_k^{-1}\left(\widetilde{A}_k x_k + \widetilde{B}_k \widetilde{K}_k \epsilon_k - (\widetilde{A}_k + \widetilde{B}_k \widetilde{K}_k)\epsilon_k\right) \\
&- T_k^{-1}\Phi_k L\left(\sum_{j \in \mathcal{N}_k} ([\widetilde{C}_j \; 0]\epsilon_j - [\widetilde{C}_k \; 0]\epsilon_k) - y_k\right) \\
=& T_k^{-1}\widetilde{A}_k T s_k + \begin{bmatrix} I_\nu \\ 0 \end{bmatrix} L \left(\sum_{j \in \mathcal{N}_k} ([\widetilde{C}_j \; 0]T_j s_j - [\widetilde{C}_k \; 0]T_k s_k)\right).
\end{aligned}$$

Using the notation $s_k = \begin{bmatrix} s_{k,1} \\ s_{k,2} \end{bmatrix}$, where $s_{k,1} \in \mathbb{R}^\nu$ and $s_{k,2} \in \mathbb{R}^{n_k-\eta_k}$ we get

$$\begin{aligned}
\dot{s}_k =& \begin{bmatrix} S & * \\ 0 & \widetilde{Q}_k \end{bmatrix} s_k + \begin{bmatrix} I_\nu \\ 0 \end{bmatrix} LR\left(\sum_{j \in \mathcal{N}_k} (s_{j,1} - s_{k,1})\right) \\
&+ \begin{bmatrix} I_\nu \\ 0 \end{bmatrix} L\left(\sum_{j \in \mathcal{N}_k} ([\widetilde{C}_j \; 0]\Phi_j^\perp s_{j,2} - [\widetilde{C}_k \; 0]\Phi_k^\perp s_{k,2})\right),
\end{aligned}$$

where (4.5) is applied in order to cancel the output heterogeneity. Since $\widetilde{Q}_k$ is a Hurwitz matrix, $s_{k,2}$ converges to zero for all $k = 1, ..., N$ and $t \to \infty$. Synchronization of $s_{k,1}$ follows with Lemma 4.2, i.e. $(s_{k,1} - s_{j,1}) \to 0$ for all k, j as $t \to \infty$. Then, synchronization of s_k leads to convergence of ϵ_k to zero:

$$\dot{\epsilon}_k = \underbrace{(\widetilde{A}_k + \widetilde{B}_k \widetilde{K}_k)}_{\text{Hurwitz matrix}} \epsilon_k + \Phi_k L \underbrace{\left(\sum_{j \in \mathcal{N}_k} ([\widetilde{C}_j \; 0]\epsilon_j - [\widetilde{C}_k \; 0]\epsilon_k) - y_k\right)}_{= -\sum_{j \in \mathcal{N}_k} ([\widetilde{C}_j \; 0]T_j s_j - [\widetilde{C}_k \; 0]T_k s_k) \to 0}$$

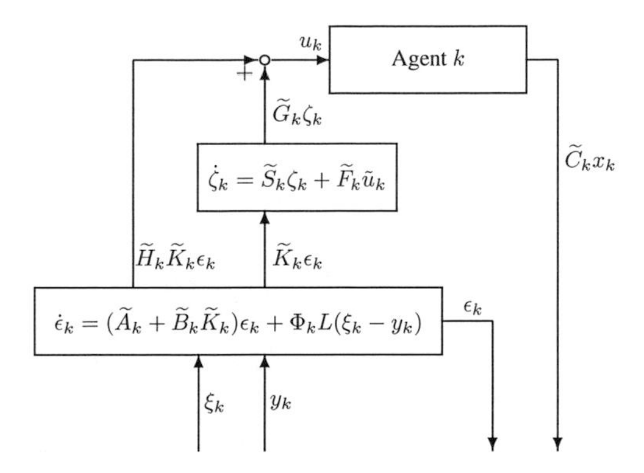

Figure 4.1: Local control structure. The arrows going in and out represent the interconnection with the other agents.

Finally, with (4.25) we get $\tilde{x}_k = T_k s_k + \epsilon_k$ for all $k = 1, ..., N$, where ϵ_k and $s_{k,2}$ converge to 0 for $t \to \infty$. Therefore, we can conclude that $[\widetilde{C}_k\ 0]\tilde{x}_k \to Rs_{k,1}$ for $t \to \infty$ which results in synchronization of z_k, $k = 1, ..., N$. ■

The local control structure given by (4.17) is depicted in Figure 4.1. Note that every controller only uses the relative outputs y_k and the relative controller states ξ_k, which can be implemented due to Assumption 4.1.

For the design of the controllers, some overall information about the communication topology is needed, as the conditions of Lemma 4.1 need to be satisfied and the algebraic Riccati equation uses $\lambda_2(\mathcal{L}_\mathcal{G})$. This kind of requirement for overall information is in alignment with most publications on synchronization of heterogeneous MAS based on output feedback, such as (Wieland, Sepulchre & Allgöwer, 2011; H. Kim et al., 2011; Seyboth et al., 2012; Seyboth & Allgöwer, 2014).

In our control scheme, a special case arises if all agents have the identical persistent components, i.e. the Regulator Equations (4.5) are satisfied with $\Gamma_k = 0$ and $S \in \mathbb{R}^{\nu\times\nu}$ with $\nu = \eta_k$, for all $k \in \mathcal{N}$. Then, the feedforward controller $\widetilde{G}_k\zeta_k$ is not needed any more and the dynamic output feedback (4.17) simplifies to

$$\begin{aligned}\dot{\epsilon}_k &= \left(\widetilde{A}_k + B_k\widetilde{K}_k\right)\epsilon_k + \Pi_k L(\xi_k - y_k)\\ u_k &= \widetilde{H}_k\widetilde{K}_k\epsilon_k \\ \xi_k &= \sum_{j\in\mathcal{N}_k}([\widetilde{C}_j\ 0]\epsilon_j - [\widetilde{C}_k\ 0]\epsilon_k).\end{aligned} \tag{4.26}$$

The control scheme (4.26) is an extension of the results from (Fax & Murray, 2004) in two ways:

(i) Extension from relative state feedback to relative output feedback.

(ii) Incorporating individual state transformation and addition of individual transient dynamics $\widetilde{Q}_k$ in (4.11).

Moreover, (4.26) can be related to the control scheme proposed in (Scardovi & Sepulchre, 2009): In (Scardovi & Sepulchre, 2009) relative states estimates are used for feedback through the application of local estimators. In contrast to (Fax & Murray, 2004), dynamic feedback as in (4.26) is applied in order to enable synchronization in the presence of a time-varying communication topology. Thus, in the case when all $A_k = A \in \mathbb{R}^{n\times n}$ are identical and all $\widetilde{C}_k = I_n$, then (4.26) yields the same result as (Scardovi & Sepulchre, 2009).

Generally, the resulting output trajectories z_k converge to the output trajectory z_0 generated by the system

$$\begin{aligned} \dot{x}_0 &= Sx_0 \\ z_0 &= Rx_0 \end{aligned}$$

for some initial condition $x_0(0)$, in the sense that $z_k(t) - z_0(t) \to 0$ for all $k \in \mathcal{N}$ and $t \to \infty$. What is special about the control scheme (4.17) is that by building $\sigma(S)$ as the union of the agents' persistent components, the resulting synchronized behaviour does not suppress any of the agents' eigenvalues. The geometric properties of this discussion in terms of invariant subspaces are shown in the following remark and simulation examples will be shown later in this chapter.

Remark 4.1 *Let* $\mathbf{x} = [\tilde{x}_j]_{j\in\mathcal{N}} \in \mathbb{R}^\Omega$ *be the stacked vector of all* $\tilde{x}_k$*, i.e.* $\Omega = \sum_{k=1}^N (n_k + \nu - \eta_k)$*. We define the subspace*

$$\mathcal{X} = \{\mathbf{x} \in \mathbb{R}^\Omega : \ \widetilde{C}_j x_j - \widetilde{C}_k x_k = 0 \, \textit{for all } j,k \in \mathcal{N}\}$$

and the supremal subspace $\mathcal{X}^* \subset \mathcal{X}$*, which is invariant with respect to the agent models (4.24). If (4.12) - (4.14) hold, then we know from (4.15) that* S *includes the persistent component of each agent (4.1) in* $\mathcal{X}^*$*. In particular, (4.16) guarantees that* $\sigma(\widetilde{A}_k) \cap j\mathbb{R}$ *are identical, and can fully be be represented by* S*. Moreover, conditions (4.5), (4.12) and (4.14) guarantee that invariance of* $\mathcal{X}^*$ *can be achieved by a feedforward controller* $u_k = \widetilde{G}_k \zeta_k$ *with* $\dot{\zeta}_k = \widetilde{S}_k \zeta_k + \widetilde{F}_k \tilde{u}_k$ *for all* $k \in \mathcal{N}$*.*

4.1.2 Synchronization by distributed reference tracking

In the last section, we discussed an approach where the resulting synchronized trajectories depend on the union of the agents' persistent components. This is in contrast to the approach proposed in (Wieland, Sepulchre & Allgöwer, 2011), where it is shown that in the case when the state x_k is available for local feedback control and given Assumption 4.2, solvability of the Francis Equations (4.5) alone is sufficient for synchronization by applying a tracking controller.

As in the previous section, exponential divergence of the agents' states is undesirable, which is why we request that

$$(S, R) \text{ is observable and } \sigma(S) \subset i\mathbb{R}. \tag{4.27}$$

Note that Assumption (4.3) is not needed here. This tracking controller is given in the following Theorem.

Theorem 4.2 *(Wieland, Sepulchre & Allgöwer, 2011) Let a group of N agents be given as* (4.1) *with synchronization outputs* (4.2) *and relative output sensing* (4.4)*, and let the communication topology be defined by a connected graph $\mathcal{G} = (\mathcal{V}, \mathcal{E})$. Let Assumption 4.1 and 4.2 be satisfied and suppose there exist a positiv integer ν and matrices $S \in \mathbb{R}^{\nu\times\nu}$, $R \in \mathbb{R}^{\tilde{r}\times\nu}$, $\Pi_k \in \mathbb{R}^{n_k\times\nu}$, $\Lambda_k \in \mathbb{R}^{m_k\times\nu}$ such that (4.5),*(4.27) *are satisfied for all $k \in \mathcal{N}$.*

Then, the dynamic controller

$$\begin{aligned} \dot{\zeta}_k &= S\zeta_k + \sum_{j\in\mathcal{N}_k} (\zeta_j - \zeta_k) \\ u_k &= K_k(x_k - \Pi_k\zeta_k) + \Gamma_k\zeta_k \end{aligned} \tag{4.28}$$

achieves synchronization if $A_k + B_kK_k$ is Hurwitz.

This control law has the advantage that it provides for degrees of freedom in the choice of the solution (S, R). This essentially means that the trajectories, which the agents outputs z_k converge to, are assigned by the implementation of a virtual reference $\dot{\zeta}_k = S\zeta_k + \delta_k$ in each local controller, where δ_k represents the cooperation between the controllers in the form of the diffusive term $\sum_{j\in\mathcal{N}_k}(\zeta_j - \zeta_k)$. By implementing δ_k in this way, the respective reference states $\zeta_k(t)$ converge towards each other and thus negotiate a common reference trajectory. Therefore, the resulting trajectory is determined through the initialization of $\zeta_k(0)$. This control scheme differs from leader-follower synchronization as every agent implements its own virtual reference, where in leader-follower systems, one reference is implemented for the whole MAS.

The local control law u_k in (4.28) aims at $x_k(t) \to \Pi_k\zeta_k(t)$ for $t \to \infty$, which means that it is a tracking controller that can be implemented in purely decentralized fashion. It should be noted that the resulting trajectories of the virtual exosystem states ζ_k are not affected by the agent states x_k, and through the tracking control law u_k, the dynamical behaviour of the unforced agent $\dot{x}_k = A_k x_k$ can be widely suppressed.

The discussion above are based on the assumption that the agents' state x_k is available for feedback control. Generally, the states are not available and therefore, enabling the dynamic controllers (4.28) substantially depends upon suitable observer design, which will be the subject of the upcoming section. But before delving into observer design, it is important to note some basic observability properties of relative output sensing (4.4), which give information about when the tracking-based approach can be used and when the model expansion approach should be used in order to address the synchronization problem subject to relative output sensing.

In fact, for analysis of observability, it is reasonable to consider the complete MAS as a relative sensing network (RSN), meaning that all measured outputs $z_j - z_k$ for $j \in \mathcal{N}_k$ are collected and then, it is considered, whether the MAS is observable with respect to this set of measurements.

The RSN can be written as

$$\dot{x} = \begin{bmatrix} A_1 & & \\ & \ddots & \\ & & A_N \end{bmatrix} x, \qquad y = \left(E_{\mathcal{G}}^\top \otimes I_{\tilde{r}}\right) \begin{bmatrix} \widetilde{C}_1 & & \\ & \ddots & \\ & & \widetilde{C}_N \end{bmatrix} x, \tag{4.29}$$

where $E_{\mathcal{G}}$ is the graph incidence matrix. Then, observability of the RSN can be well analyzed using results presented in (Zelazo & Mesbahi, 2008, 2011), where we summarize the relevant part below:

Theorem 4.3 *(Zelazo & Mesbahi, 2011) The heterogeneous relative sensing network* (4.29) *is unobservable if and only if the following conditions are met:*

1. *There exists an eigenvalue λ that is common to each $A_k, k \in \mathcal{N}$.*

2. *For all $k, j \in \mathcal{N}$, $A_k x_k = \lambda x_k$ and $A_j x_j = \lambda x_j$ implies that $\widetilde{C}_k x_k = \widetilde{C}_j x_j$.*

This theorem is a valuable result, which essentially says that in the case of relative output sensing as considered in this section, observer design for applying (4.28) can only be applied if there is no common eigenvalue of all agents, whose eigenvector produces the identical output for every agent. The next section will specifically deal with observer design for enabling (4.28), so one should keep in mind that the following results cannot always be applied: Instead, given certain circumstances, the approach discussed in Section 4.1.1 is more suitable to achieve synchronization.

4.2 Regional Estimation-based Synchronization

In the previous section, we considered nominal systems without disturbances in order to establish basic geometric properties and present two approaches to synchronization. In the following, we consider the extended problem where in addition to synchronization in the nominal case, we seek performance guarantees for the synchronizing behaviour with respect to exogeneous disturbances.

Let a group of N LTI agents be described by the differential equation

$$\dot{x}_k = A_k x_k + B_k u_k + B_k^w w_k, \tag{4.30}$$

where for all $k \in \mathcal{N}$, $x_k(t) \in \mathbb{R}^{n_k}$ is the state of the agent, $u_k(t) \in \mathbb{R}^{m_k}$ is the respective control input, and w_k with $w_k(t) \in \mathbb{R}^{v_k}$ is the $\mathcal{L}_2$-integrable disturbance affecting agent k.

The outputs to be synchronized are again defined by (4.2) and the communication topology between the agents is defined by the directed graph $\mathcal{G} = (\mathcal{V}, \mathcal{E})$, i.e. agent k may receive information from agent j if $(v_j, v_k) \in \mathcal{E}$.

Concerning the measured outputs y_k, we introduce the more general formulation of

$$y_k = C_k x_k + \sum_{j \in \mathcal{N}_k} C_{kj} x_j + \eta_k, \tag{4.31}$$

where η_k with $\eta_k(t) \in \mathbb{R}^{r_k}$ is a $\mathcal{L}_2$-integrable measurement disturbance function. Besides the case of pure relative output sensing (4.4), this formulation also includes the absolute sensing case of $y_k = z_k, k \in \mathcal{N}$ or hybrid cases where $y_k = z_k, k \in \mathcal{I}$ and (4.4) for $k \in \mathcal{N} \backslash \mathcal{I}$, to name a few. Assumptions 4.1 and 4.2 are used again as basic properties.

With the definition of the system class under consideration and the disagreement function

$$\Phi = \sum_{k=1}^{N} \sum_{j \in \mathcal{N}_k} \|z_k - z_j\|^2 \tag{4.32}$$

we can now propose the refined closed-loop synchronization problem.

Problem 5 (Robust closed loop output synchronization) *For every agent k, determine a dynamic controller, which receives the measurement y_k and data communicated from the neighbouring agents $j \in \mathcal{N}_k$, such that:*

1. *In the absence of exogenous and measurement disturbances (i.e., when* $w_k = 0$*,* $\eta_k = 0$*, for all* $k \in \mathcal{N}$*) synchronization is achieved.*

2. $\mathcal{H}_\infty$*-type synchronization performance is guaranteed in the sense that*

$$\int_0^\infty \Phi(t)dt \leq \sum_{k=1}^N \int_0^\infty (\kappa^2\|w_k\|^2 + \theta^2\|\eta_k\|^2)dt + \widetilde{I}_0 \tag{4.33}$$

with performance parameters $\kappa, \theta > 0$ *and the initial costs* $\widetilde{I}_0 \geq 0$ *due to the transient synchronization error, which depends on the initialization of the agents and the dynamic controllers.*

Following the discussions from Section 4.1.2, in order to solve Problem 5, we aim for an observer-based approach, where the observers assigned to the agents estimate the agents' states in a distributed fashion, and then, a synchronizing controller such as (4.28) is applied. Therefore, the next step is to define and address the corresponding observer design problem that subsequently enables the synchronizing controller.

4.2.1 Regional Estimation Scheme

First, we elaborate on the basic detectability properties of the multi-agent system. With the stacked vectors $x = [x_k]_{k\in\mathcal{N}}$, $u = [u_k]_{k\in\mathcal{N}}$, $w = [w_k]_{k\in\mathcal{N}}$, $\eta = [\eta_k]_{k\in\mathcal{N}}$, $y = [y]_{k\in\mathcal{N}}$, the global system can be written as

$$\dot{x} = \text{diag}[A_k]_{k\in\mathcal{N}}x + \text{diag}[B_k]_{k\in\mathcal{N}}u + \text{diag}[B_k^w]_{k\in\mathcal{N}}w \tag{4.34}$$

with the global output

$$y = \underbrace{\begin{bmatrix} C_1 & C_{12} & \dots & C_{1N} \\ C_{21} & C_2 & & \vdots \\ \vdots & & \ddots & \\ C_{N1} & \dots & & C_N \end{bmatrix}}_{C} x + \eta. \tag{4.35}$$

In order for the estimation problem to be solvable, we necessarily need global detectability.

Assumption 4.4 *The pair* (A, C) *is detectable.*

With the basic detectability property satisfied, we have two intuitive approaches for distributed observer design, which both have significant disadvantages and thus motivate the usage of our scheme that we are proposing afterwards.

Local estimation: The first intuitive approach can be found by designing every observer to only estimate the state of its respective agent x_k, while receiving the estimates of the neighbouring agents $\hat{x}_j$. Thus, the observers can be written as

$$\dot{\hat{x}}_k = A_k\hat{x}_k + \sum_{j\in\mathcal{N}_k} A_{kj}\hat{x}_j + L_k(y_k - C_k\hat{x}_k - \sum_{j\in\mathcal{N}_k} C_{kj}\hat{x}_j) + B_k u_k. \tag{4.36}$$

As a matter of fact, this approach is similar to the centralized Luenberger observer (A.6). However, while the centralized Luenberger observer allows for a full-block observer gain matrix L, the decentralized structure of (4.36) requires that L is block-diagonal. While this may work in some examples (Langbort & Ugrinovskii, 2010), it is restrictive in general. One example, where this intuitive approach fails is given later in this chapter. Moreover, this setup requires constant communication between the observers.

Global estimation: The second intuitive approach can be found by directly applying the observers (2.6) as already pointed out at the beginning of this chapter:

$$\dot{\hat{x}}_k = Ax_k + L_k(y_k - C_k\hat{x}_{k,k} - \sum_{j\in\mathcal{N}_k} C_{kj}\hat{x}_{k,j}) + K_k \sum_{j\in\mathcal{N}_k}(x_j - x_k) + Bu. \tag{4.37}$$

Here, A, B are the agglomerated matrices defined in (4.34). The second index j in $x_{k,j}$ is used to indicate the component of the estimate x_k with respect to agent j. This estimation scheme has shown to be less restrictive than (4.36), however leads to every observer creating an estimate of the global system, which is undesirable as the complexity of every individual estimator grows with the number of agents, and all control laws u_k, $k \in \mathcal{N}$ need to be gathered. Applying the scheme from (Olfati-Saber, 2005), (Olfati-Saber, 2006) leads to the same effect.

The two approaches for observer design introduced above (4.36),(4.37), either allow the individual observers to only estimate its corresponding agent state x_k or to estimate the complete system state x including all agents. The main idea behind the estimation scheme that we are proposing in this section is based on the middle ground between these two intuitive observer design approaches: We propose to implement *regional estimation*, i.e. the observer k estimate the state of its corresponding agent x_k and additionally the states of all neighbouring agents $x_j, j \in \mathcal{N}_k$.

Thus, with $\mathcal{N}_k = \{j_1, ..., j_{p_k}\}$, let the vector of local estimates be defined as

$$\hat{x}_k = \begin{bmatrix} \hat{x}_{k,k} \\ \hat{x}_{k,j_1} \\ \vdots \\ \hat{x}_{k,j_{p_k}} \end{bmatrix} = \begin{bmatrix} \hat{x}_{k,k} \\ [\hat{x}_{k,j}]_{j \in \mathcal{N}_k} \end{bmatrix} \in \mathbb{R}^{n_k + \sigma_k} \tag{4.38}$$

and the corresponding error vector be defined as

$$e_k = \overline{x}_k - \hat{x}_k = \begin{bmatrix} x_k - \hat{x}_{k,k} \\ x_{j_1} - \hat{x}_{k,j_1} \\ \vdots \\ x_{j_{p_k}} - \hat{x}_{k,j_{p_k}} \end{bmatrix} = \begin{bmatrix} x_k - \hat{x}_{k,k} \\ [x_j - \hat{x}_{k,j}]_{j \in \mathcal{N}_k} \end{bmatrix} \in \mathbb{R}^{n_k + \sigma_k}, \tag{4.39}$$

where σ_k is the total dimension of the neighbouring agents' states. The second index j denotes the component of an estimate corresponding to agent j, and the state vector $\overline{x}_k$ including the state of the agent x_k and its neighbours $x_j, j \in \mathcal{N}_k$ will be referred to as *combined state*.

Remark 4.2 *The definition of the combined states $\overline{x}_k$ was made based on Assumption 4.1. In fact, all results in this section can also be extended to the case where the communication topology is different from the measurement topology. In this case, the construction of the combined states $\overline{x}_k$ follows the graph representing the measurement topology. However, we neglect this possible extension for the sake of simplicity and since Assumption 4.1 is very common in the literature of distributed/cooperative control, see e.g. the literature review in the introduction.*

For the purpose of observer design, we also assume for now that the observer k has knowledge about the models of the neighbouring agents $j \in \mathcal{N}_k$. Since the neighbouring agents may communicate their model, this is not restrictive in general. In some examples however, we can further relax it as shown later in this chapter.

Defining the local estimates as (4.38) allows to include the advantages of both (4.36) and (4.37): On the one hand, implementing *regional* instead of *global* estimation, the dimension of an observer only grows, if a newly added agent is a neighbour, which means that scalability of the system is preserved. And moreover, the observers may cooperate as in (4.37). With the definition of the system class under consideration, we can now state the estimation problem.

Problem 6 (Regional estimation) *For every agent k, determine an observer for the combined state $\overline{x}_k$, which receives the measurement y_k and data communicated from the neighbouring agents $j \in \mathcal{N}_k$, such that the estimate $\hat{x}_k$ satisfies:*

1. *In the absence of model and measurement disturbances (i.e., when $w_k = 0$, $\eta_k = 0$, for all $k = 1, ..., N$), the estimation errors $e_k = \overline{x}_k - \hat{x}_k$ decay so that $e_k \to 0$ exponentially for all $k \in \mathcal{N}$.*

2. *The observers provide guaranteed $\mathcal{H}_\infty$ performance in the sense that*

$$\sum_{k=1}^{N} \int_0^\infty e_k^\top W_k e_k dt \le \gamma^2 \sum_{k=1}^{N} \int_0^\infty (\|q_k w_k\|^2 + \|\eta_k\|^2) dt + I_0 \tag{4.40}$$

where W_k is a given positive semi-definite weighting matrix and I_0 is the initial cost due to the transient estimation error depending on the initial conditions of the observers.

Solving Problem 6 is an important component to solving Problem 5, as bounding the influence of the estimation error is needed to ensure closed-loop performance. In the following, we present LMI-based methods to ensure these properties.

4.2.2 Proposed observer design

In order to solve Problem 6, we need to adapt the estimation scheme from Chapter 2 to the modified task of estimating the combined state $\overline{x}_k$ instead of the global state x.

Let the observer be defined as

$$\dot{\hat{x}}_k = \overline{A}_k \hat{x}_k + L_k (y_k - \overline{C}_k \hat{x}_k) + K_k \left[\hat{x}_{j,j} - \hat{x}_{k,j} \right]_{j \in \mathcal{N}_k} + \overline{B}_k \overline{u}_k, \tag{4.41}$$

with the respective matrix definitions

$$\begin{aligned} \overline{A}_k &= \begin{bmatrix} A_k & 0 \\ 0 & \text{diag}[A_j]_{j\in\mathcal{N}_k} \end{bmatrix}, \overline{C}_k &= \begin{bmatrix} C_k & C_{kj_1} & \dots & C_{kj_{p_k}} \end{bmatrix}, \\ \overline{B}_k &= \begin{bmatrix} B_k & 0 \\ 0 & \text{diag}[B_j]_{j\in\mathcal{N}_k} \end{bmatrix}, \overline{u}_k &= \begin{bmatrix} u_k \\ [u_j]_{j\in\mathcal{N}_k} \end{bmatrix}, \end{aligned} \tag{4.42}$$

for $\mathcal{N}_k = \{j_1, ..., j_{p_k}\}$. The initial condition can be set as $\hat{x}_k(0) = 0$ and the filter gains to be designed are $L_k \in \mathbb{R}^{(n_k+\sigma_k)\times r_k}$ and $K_k \in \mathbb{R}^{(n_k+\sigma_k)\times \sigma_k}$.

Remark 4.3 *Considering Remark 4.2, we observe that if there are less communication channels than assumed in Assumption 4.1, then this results in the structural constraint that some columns of K_k are 0. This can easily be implemented.*

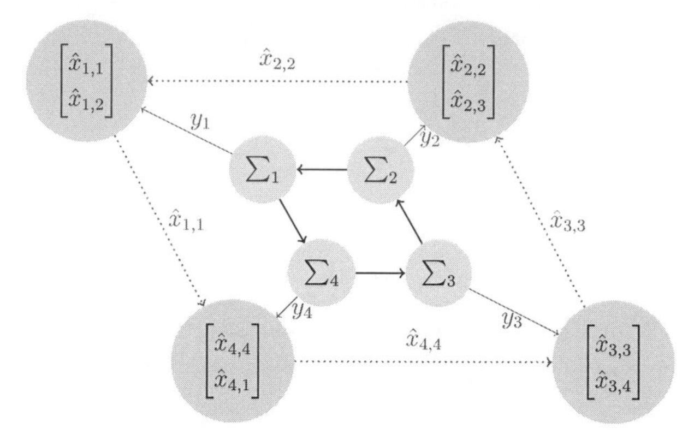

Figure 4.2: Example for the estimator structure. The inner graph represents the agents and the measurement topology, the outer graph represents the estimators and the communication topology.

The observers (4.41) are composed as follows:

$$\begin{array}{lll} \overline{A}_k\hat{x}_k & \dots & \text{Agent state feedback} \\ +L_k(y_k - \overline{C}_k\hat{x}_k) & \dots & \text{Correction based on local measurement} \\ +K_k\begin{bmatrix}\hat{x}_{j,j} - \hat{x}_{k,j}\end{bmatrix}_{j\in\mathcal{N}_k} & \dots & \text{Consensus term} \\ +\overline{B}_k\overline{u}_k & \dots & \text{Control inputs.} \end{array} \tag{4.43}$$

Remark 4.4 *Since the observers* (4.41) *incorporate estimates of the neighboring agents, naturally, they also make use of* $\overline{u}_k = \begin{bmatrix} u_k^\top & u_{j_1}^\top & \dots & u_{j_{p_k}}^\top \end{bmatrix}^\top$, *i.e. the neighbouring agents' control inputs. This may be undesirable as it increases communication load. However, in fact this can be eliminated in some cases, when the control inputs* $u_k, k = 1, ..., N$ *are defined by some control law. One example for this will be given later.*

For the example with four agents and directed ring-type topology, i.e. $(v_{k+1}, v_k) \in \mathcal{E}$ for $k = 1, ..., 3$ and $(v_1, v_4) \in \mathcal{E}$, the observer architecture can be depicted as shown in Figure 4.2 (Wu et al., 2014).

We can now particularize the design problem of this section.

Problem 6* *Determine estimator gains* L_k, K_k *such that the distributed observers* (4.41) *solve Problem 6.*

The weighting matrix W_k used in the $\mathcal{H}_\infty$−type performance (4.40), is a design parameter and can be chosen as needed. Later, it will turn out that the specific choice of W_k is determined by the synchronizing control law.

With the distributed observers (4.41) and the agents (4.30), it holds for the estimation error e_k that

$$\dot{e}_k = (\overline{A}_k - L_k\overline{C}_k)e_k + K_k \left[e_{j,j} - e_{k,j}\right]_{j\in\mathcal{N}_k} - L_k\eta_k + \overline{B}_k^w \overline{w}_k, \tag{4.44}$$

with

$$\overline{B}_k^w = \begin{bmatrix} B_k^w & 0 \\ 0 & \text{diag}[B_j^w]_{j\in\mathcal{N}_k} \end{bmatrix}, \quad \overline{w}_k = \begin{bmatrix} w_k \\ [w_j]_{j\in\mathcal{N}_k} \end{bmatrix}.$$

In analogy to (2.15), we define the matrices

$$\begin{aligned} Q_k =& P_k\overline{A}_k + \overline{A}_k^\top P_k - G_k^\top \overline{C}_k - \overline{C}_k^\top G_k - F_k \left[0_{\sigma_k\times n_k} I_{\sigma_k}\right] - \left[0_{\sigma_k\times n_k} I_{\sigma_k}\right]^\top F_k^\top \\ &+ \alpha_k P_k + q_k \begin{bmatrix} \widetilde{P}_k & 0 \\ 0 & 0 \end{bmatrix}, \end{aligned} \tag{4.45}$$

where $P_k \in \mathbb{R}^{(n_k+\sigma_k)\times(n_k+\sigma_k)}$ and $\widetilde{P}_k \in \mathbb{R}^{n_k\times n_k}$ are symmetric, positive definite matrices. F_k and G_k are $(n_k+\sigma_k)\times\sigma_k$-dimensional and $(n_k+\sigma_k)\times r_k$-dimensional solution variables, respectively, and $\alpha_k > 0$ is a positive constants which can be defined to determine convergence speed. With these definitions, we are ready to present our main result.

Theorem 4.4 *Consider a group of N agents defined by* (4.30),(4.31) *for all $k \in \mathcal{N}$ and suppose Assumption 4.1 holds. Let a collection of matrices F_k, $P_k \succ 0$, $\widetilde{P}_k \succ 0$, $k \in \mathcal{N}$, and a performance parameter $\beta \geq 0$ be a solution of the LMIs*

$$\left[\begin{array}{ccc:c} Q_k + W_k & -G_k & P_k\overline{B}_k^w & F_k \\ -G_k^\top & -\beta I & 0 & 0 \\ (P_k\overline{B}_k^w)^\top & 0 & -\beta I & 0 \\ \hdashline F_k^\top & 0 & 0 & -diag\left[\widetilde{P}_j\right]_{j\in\mathcal{N}_k} \end{array}\right] \preceq 0 \tag{4.46}$$

for all $k = 1, ..., N$, then Problem 6 admits a solution of the form (4.41) *with*

$$\begin{aligned} L_k &= (P_k)^{-1}G_k, \\ K_k &= (P_k)^{-1}F_k, \\ \gamma &= \sqrt{\beta}. \end{aligned} \tag{4.47}$$

Proof. Consider the Lyapunov function

$$V(e) = \sum_{k=1}^{N} \underbrace{e_k^\top P_k e_k}_{V_k(e_k)} .$$

For the derivative of e_k, we have (4.44) and it follows for the Lie derivative of the components $V_k(e_k)$ that

$$\dot{V}_k(e_k) = 2e_k^\top P_k(\overline{A}_k - L_k\overline{C}_k)e_k + 2e_k^\top P_k(-L_k\eta_k + \overline{B}_k^w \overline{w}_k) + 2e_k^\top P_k K_k \left[e_{j,j} - e_{k,j}\right]_{j\in\mathcal{N}_k} .$$

With the filter gains (4.47) and the definition (4.45) it can be obtained that

$$\begin{aligned}\dot{V}_k(e) =& e_k^\top Q_k e_k - e_k^\top \left(\alpha_k P_k + q_k\pi_k \begin{bmatrix} \widetilde{P}_k & 0 \\ 0 & 0 \end{bmatrix}\right) e_k \\ & - 2e_k^\top G_k^\top \eta_k + 2e_k^\top P_k \overline{B}_k^w \overline{w}_k + 2e_k^\top F_k \left[e_{j,j}\right]_{j\in\mathcal{N}_k}\end{aligned}$$

and with the LMIs (4.46), we have

$$\dot{V}_k(e) \leq \sum_{j\in\mathcal{N}_k} \pi_j e_{j,j}^\top P_{11}^j e_{j,j} - e_k^\top W_k e_k + \beta\eta_k^\top \eta_k + \beta\overline{w}_k^\top \overline{w}_k - \alpha_k e_k^\top P_k e_k - q_k\pi_k e_{k,k}^\top \widetilde{P}_k e_{k,k}.$$

Summing up the V_ks, it holds for V that

$$\begin{aligned}\dot{V}(e) \leq & \underbrace{\sum_{k=1}^{N}\sum_{j\in\mathcal{N}_k} \pi_j e_{j,j}^\top P_{11}^j e_{j,j}}_{=\sum_{k=1}^N q_k\pi_k e_{k,k}^\top \widetilde{P}_k e_{k,k}} - \sum_{k=1}^{N} e_k^\top W_k e_k + \sum_{k=1}^{N} \beta\eta_k^\top \eta_k + \sum_{k=1}^{N} \beta\overline{w}_k^\top \overline{w}_k \\ & - \sum_{k=1}^{N} \alpha_k e_k^\top P_k e_k - q_k\pi_k \sum_{k=1}^{N} e_{k,k}^\top \widetilde{P}_k e_{k,k} \\ = & - \sum_{k=1}^{N} \alpha_k \underbrace{e_k^\top P_k e_k}_{V_k} - \sum_{k=1}^{N} e_k^\top W_k e_k + \sum_{k=1}^{N} \beta\eta_k^\top \eta_k + \sum_{k=1}^{N} \beta\overline{w}_k^\top \overline{w}_k\end{aligned} \tag{4.48}$$

Through integration on the interval $[0,T]$, we obtain

$$\begin{aligned}V(e(T)) + \sum_{k=1}^{N}\int_0^T e_k^\top W_k e_k dt \leq & \beta \sum_{k=1}^{N}\int_0^T (\|\overline{w}_k\|^2 + \|\eta_k\|^2)dt + \sum_{k=1}^{N} e_k^\top(0)P_k e_k(0) \\ \leq & \beta\cdot \sum_{k=1}^{N}\int_0^T (q_k\|w_k\|^2 + \|\eta_k\|^2)dt + \sum_{k=1}^{N} e_k^\top(0)P_k e_k(0).\end{aligned}$$

As $V(e(T)) \geq 0$ and with the zero initial conditions of the observer states, it follows that

$$\sum_{k=1}^{N} \int_0^T e_k^\top W_k e_k dt \leq \gamma^2 \sum_{k=1}^{N} \int_0^T (q_k \|w_k\|^2 + \|\eta_k\|^2) dt + I_0.$$

Now, let $T \to \infty$, then this satisfies Property 2 of Problem 6. Moreover, if $w_k = 0$ and $\eta_k = 0$ for all $k \in \mathcal{N}$, then it follows from (4.48) that

$$\dot{V}(e) \leq -\min_k(\alpha_k) V,$$

which implies that Property 1 of Problem 6 holds. ■

As it will be shown in the next section, the choice of W_k directly depends upon the gain matrices of the synchronizing controller, in order to show closed-loop synchronization performance as required by Problem 5.

Remark 4.5 *The choice of α_k determines the convergence speed of the estimators, where a larger α_k enforces faster convergence of the estimates, but also requires larger filter gains in general.*

4.2.3 Robustly synchronizing controllers

In this section we propose two synchronizing controllers, where the first one directly employs the approach from Theorem 4.2. Suppose the communication graph $\mathcal{G}$ is connected, then the first controller is proposed as

$$\dot{\zeta}_k = S\zeta_k + \underbrace{\sum_{j \in \mathcal{N}_k} (\zeta_j - \zeta_k)}_{\delta_k} \tag{4.49}$$

$$u_k = \Gamma_k \zeta_k + H_k(\hat{x}_{k,k} - \Pi_k \zeta_k), \tag{4.50}$$

where the matrices $S \in \mathbb{R}^{\nu \times \nu}, R \in \mathbb{R}^{\tilde{r} \times \nu}, \Pi_k \in \mathbb{R}^{n_k \times \nu}$ and $\Gamma_k \in \mathbb{R}^{m_k \times \nu}$ are obtained from the Francis equations (4.5) with $S \in \mathbb{R}^{\nu \times \nu}$.

As discussed in Section 4.1, solvability of (4.5) with (S, R) being observable is a necessary condition for synchronization of the nominal systems (4.30). Therefore, the existence of a solution to (4.5) is not conservative and moreover, it is commonly assumed in the literature that the agents know of this solution.

In the following, the term $\epsilon_k = x_k - \Pi_k \zeta_k$ will be called the local regulation error. Note that if $\epsilon_k \to 0$ for all $k \in \mathcal{N}$ and in addition, $\zeta_k - \zeta_j \to 0$ for all $k, j \in \mathcal{N}$, then we have asymptotic synchronization because

$$z_k - z_j = \widetilde{C}_k \epsilon_k - \widetilde{C}_j \epsilon_j + R\zeta_k - R\zeta_j$$

for all $j, k \in \mathcal{N}$. More specifically, the disagreement function Φ (4.32) can be upper bounded using the following result.

Lemma 4.3 *Suppose we have exponential convergence of $\zeta_k - \zeta_j \to 0$ for all $k, j \in \mathcal{N}$. Then, the inequality*

$$\sum_{k=1}^{N} \sum_{j \in \mathcal{N}_k} \int_0^\infty \|\widetilde{C}_k \epsilon_k - \widetilde{C}_j \epsilon_j\|^2 dt \leq \sum_{k=1}^{N} \int_0^\infty (\overline{\kappa}^2 \|w_k\|^2 + \overline{\theta}^2 \|\eta_k\|^2) dt + \overline{I}_0 \tag{4.51}$$

with the parameters $\overline{\kappa}, \overline{\theta} > 0$, and the initial costs $\overline{I}_0 \geq 0$ implies that for every set of performance parameters $\kappa > \overline{\kappa}, \theta > \overline{\theta}$, there exists $\widetilde{I}_0 \geq 0$ such that (4.33) *is satisfied.*

Proof. The disagreement function Φ can be upper bounded by

$$\begin{aligned} \Phi &= \sum_{k=1}^{N} \sum_{j \in \mathcal{N}_k} \|\widetilde{C}_k \epsilon_k - \widetilde{C}_j \epsilon_j + R\zeta_k - R\zeta_j\|^2 \\ &\leq (1 + \frac{1}{\omega}) \sum_{k=1}^{N} \sum_{j \in \mathcal{N}_k} \|\widetilde{C}_k \epsilon_k - \widetilde{C}_j \epsilon_j\|^2 + (1+\omega) \underbrace{\sum_{k=1}^{N} \sum_{j \in \mathcal{N}_k} \|R\zeta_k - R\zeta_j\|^2}_{\Xi}, \end{aligned} \tag{4.52}$$

with an arbitrary $\omega > 0$. Moreover, we know from (Scardovi & Sepulchre, 2009), that indeed $\zeta_j - \zeta_k \to 0$ exponentially for all $k, j \in \mathcal{N}$. Therefore, $\int_0^\infty \Xi(t) dt$ is finite and with (4.51) it holds that

$$\begin{aligned} \int_0^\infty \Phi(t) dt &\leq \int_0^\infty (1 + \frac{1}{\omega}) \sum_{k=1}^{N} \sum_{j \in \mathcal{N}_k} \|\widetilde{C}_k \epsilon_k - \widetilde{C}_j \epsilon_j\|^2 dt + \int_0^\infty \Xi(t) dt \\ &\leq (1 + \frac{1}{\omega}) \left(\sum_{k=1}^{N} \int_0^\infty (\overline{\kappa}^2 \|w_k\|^2 + \overline{\theta}^2 \|\eta_k\|^2) dt + \overline{I}_0 \right) + \int_0^\infty \Xi(t) dt \end{aligned}$$

Therefore, (4.33) holds with $\kappa = (1 + \frac{1}{\omega})\overline{\kappa}$, $\theta = (1 + \frac{1}{\omega})\overline{\theta}$ for any $\omega > 0$. ■

Lemma 4.3 shows that (4.51) is a suitable replacement for (4.33). Furthermore, the initial costs $\widetilde{I}_0$ in (4.33) can now be specifically explained as the transient synchronization error that is caused by three separate parts:

1. The synchronization error of the local references (4.49), where large initial mismatches between ζ_k cause large $\widetilde{I}_0 \geq 0$ in general.

2. The estimation error (4.44) due to the observers uncertainty about the agents' initial condition.

3. The initial regulation error ϵ_k that needs to be stabilized through observer-based state feedback.

Item 1 vanishes when ζ_k, $k \in \mathcal{N}$, are initialized identically, which equals the case of a global reference trajectory that is available to every agent. The technical significance of the approach (4.49) lies in the fact, that in praxis it may not be possible to initialize all ζ_k exactly identical and therefore the agents must agree upon a common trajectory in a cooperative way.

A solution to Problem 5 is given in the following Theorem.

Theorem 4.5 *Let a group of N agents be given as* (4.30) *with synchronization outputs* (4.2) *and measurement outputs* (4.31)*, and let the communication topology be defined by a connected graph $\mathcal{G} = (\mathcal{V}, \mathcal{E})$. Let Assumption 4.1 and 4.2 be satisfied and suppose the following conditions hold:*

1. There exists a positiv integer ν and matrices $S \in \mathbb{R}^{\nu\times\nu}$, $R \in \mathbb{R}^{\widetilde{r}\times\nu}$, $\Pi_k \in \mathbb{R}^{n_k\times\nu}$, $\Gamma_k \in \mathbb{R}^{m_k\times\nu}$ such that (4.5),(4.27) are satisfied for all $k \in \mathcal{N}$.

2. There exist constants $\mu > 0$ and $\lambda > 0$ such that for each $k \in \mathcal{N}$, the algebraic Riccati equation

$$X_k A_k + A_k^\top X_k + 2(q_k + p_k)\widetilde{C}_k^\top \widetilde{C}_k - X_k \left(\frac{1}{\lambda^2} B_k B_k^\top - \frac{1}{\mu^2} B_k^w B_k^{w\top} - \frac{1}{\mu^2} \Pi_k \Pi_k^\top \right) X_k = 0 \tag{4.53}$$

has a positive definite symmetric solution X_k such that $A_k - \frac{1}{\lambda^2} B_k B_k' X_k$ is a Hurwitz matrix.

3. The observers (4.41) *are designed such that they solve Problem 6 with the weighting matrices*

$$W_k = \begin{bmatrix} \frac{1}{\lambda^2} X_k B_k B_k^\top X_k & 0 \\ 0 & 0 \end{bmatrix} \tag{4.54}$$

and $\gamma > 0$.

Then the control law, given as (4.50) *with*

$$H_k = -\frac{1}{\lambda^2} B_k^\top X_k, \tag{4.55}$$

solves Problem 5 with any $\kappa^2 > \mu^2 + q_{max}\gamma^2$ *and* $\theta^2 > \gamma^2$, *where* $q_{max} = \max_{k \in \mathcal{N}} q_k$.

Proof. This proof is shown in Appendix B.1 ■

Theorem 4.5 gives three item required for designing the synchronizing controllers. Thus, this represents a three step approach to solve Problem 5. Some remarks on the choice of the parameters λ, μ are in order:

Remark 4.6 *The parameters* $\lambda, \mu > 0$ *prescribe the regulator's sensitivity with respect to disturbances and the estimation error. In particular,* μ *is a lower bound for* κ *and should therefore be chosen preferably small. Depending on* μ, λ *also needs to be chosen small in order to preserve the feasibility of* (4.53). *For instance, if* $B_k = B_k^w$ *for all* $k \in \mathcal{N}$, *then* $\lambda < \mu$ *is needed. Standard Riccati-solvers such as* `care` *in* `MATLAB` *may be used even though* $\frac{1}{\lambda^2} B_k B_k^\top - \frac{1}{\mu^2} B_k^w B_k^{w\top} - \frac{1}{\mu^2} \Pi_k \Pi_k^\top$ *is indefinite. However, it has to be ensured manually that* $A_k - \frac{1}{\lambda^2} B_k B_k^\top X_k$ *is Hurwitz for all* $k \in \mathcal{N}$.

Concerning the computation needed for solving these three steps, we have following characteristics:

1. The solution of the regulator equations (4.5) is assumed to be known to all agents. This reflects the fact, that the agents need to know what kind of trajectory they will agree upon.

2. Every agents solves the Ricatti equation (4.53). The calculation can be done completely decentralized as it only includes model parameters and solution variables X_k of the individual agent. The parameters λ and μ may be agreed upon offline or using a consensus algorithm.

3. The LMIs (4.46) used for observer design are coupled in their solution variables. Therefore, the solution cannot be found in a purely decentralized manner. Distributed solution of the LMIs can however be enabled and an approach to achieve this can be adopted from the results presented in Chapter 3.

The decentralized character of Step 2 is a great advantage, which however comes with the expense of potentially weakened performance guarantees on the closed-loop behaviour. In order to improve the performance that can be achieved, we extend the synchronizing controller (4.50) by additional coupling terms. The extended controller is then proposed as

$$\dot{\zeta}_k = S\zeta_k + \underbrace{\sum_{j\in\mathcal{N}_k}(\zeta_j - \zeta_k)}_{\delta_k} \tag{4.56}$$

$$u_k = \Gamma_k\zeta_k + H_k(\hat{x}_{k,k} - \Pi_k\zeta_k) + \sum_{j\in\mathcal{N}_k} H_{kj}(\widetilde{C}_j\hat{x}_{k,j} - R\zeta_k), \tag{4.57}$$

where the matrices $S \in \mathbb{R}^{\nu\times\nu}, R \in \mathbb{R}^{\tilde{r}\times\nu}, \Pi_k \in \mathbb{R}^{n_k\times\nu}$ and $\Gamma_k \in \mathbb{R}^{m_k\times\nu}$ are obtained from the Regulator equations (4.5). The local regulation error is again defined as $\epsilon_k = x_k - \Pi_k\zeta_k$. With this controller, we have following solution to Problem 5.

Theorem 4.6 *Let a group of N agents be given as* (4.30) *with synchronization outputs* (4.2) *and measurement outputs* (4.31)*, and let the communication topology be defined by a connected graph $\mathcal{G} = (\mathcal{V}, \mathcal{E})$. Let Assumption 4.1 and 4.2 be satisfied and suppose the following conditions hold:*

1. *There exists a positiv integer ν and matrices $S \in \mathbb{R}^{\nu\times\nu}$, $R \in \mathbb{R}^{\tilde{r}\times\nu}$, $\Pi_k \in \mathbb{R}^{n_k\times\nu}$, $\Gamma_k \in \mathbb{R}^{m_k\times\nu}$ such that (4.5),(4.27) are satisfied for all $k \in \mathcal{N}$.*

2. *Let $\lambda, \mu, \delta > 0$. For all $k = 1, ..., N$, let $\mathcal{N}_k = \{j_1, ..., j_{p_k}\}$ and let the collection of matrices $X_k \succ 0$, $\widetilde{X}_k \succ 0$, $H_{kj_1}, ..., H_{kj_{p_k}}$ be a solution of the two LMIs*

$$\begin{bmatrix} \widetilde{Q}_k + \widetilde{X}_k & X_k\widetilde{C}_k^\top \\ \widetilde{C}_kX_k & -\frac{1}{(1+\delta)q_k+p_k}I_r \end{bmatrix} \preceq 0 \tag{4.58}$$

$$\begin{bmatrix} -\widetilde{X}_k & B_kH_{kj_1} - X_k\widetilde{C}_k^\top & \dots & B_kH_{kj_{p_k}} - X_k\widetilde{C}_k^\top \\ * & -\delta I_r & & \\ \vdots & & \ddots & \\ * & & & -\delta I_r \end{bmatrix} \preceq 0, \tag{4.59}$$

with $\widetilde{Q}_k = A_kX_k + X_kA_k^\top - \frac{1}{\lambda^2}B_kB_k^\top + \frac{1}{\mu^2}(B_k^wB_k^{w\top} + \Pi_k\Pi_k^\top + B_kB_k^\top) + \widetilde{\alpha}_kX_k$

and arbitrarily small $\widetilde{\alpha}_k > 0$.

3. *The observers* (4.41) *are designed such that they solve Problem 6 with the weighting matrices*

$$W_k = \begin{bmatrix} \frac{1}{\lambda}X_k^{-1\top}B_k \\ \lambda\widetilde{C}_{j_1}^\top H_{kj_1}^\top \\ \vdots \\ \lambda\widetilde{C}_{j_{p_k}}^\top H_{kj_{p_k}}^\top \end{bmatrix} \begin{bmatrix} \frac{1}{\lambda}X_k^{-1\top}B_k \\ \lambda\widetilde{C}_{j_1}^\top H_{kj_1}^\top \\ \vdots \\ \lambda\widetilde{C}_{j_{p_k}}^\top H_{kj_{p_k}}^\top \end{bmatrix}^\top \tag{4.60}$$

and the performance parameter $\gamma > 0$.

Then the control law (4.57) *with* $H_k = -\frac{1}{\lambda^2} B_k^\top X_k^{-1}$ *and the solution matrices* H_{kj} *solves Problem* (5) *with any* $\kappa^2 > \mu^2 + q_{max}\gamma^2$ *and* $\theta^2 > \gamma^2$.

Proof. Suppose the conditions of Theorem 4.6 hold.

With equations (4.56) and (4.57), the derivative of ϵ_k is

$$\begin{aligned}\dot{\epsilon}_k =&(A_k + B_k H_k)\epsilon_k + B_k \sum_{j\in\mathcal{N}_k} H_{kj}\widetilde{C}_j\epsilon_j + \underbrace{B_k \sum_{j\in\mathcal{N}_k} H_{kj}R(\zeta_j - \zeta_k)}_{\widetilde{\delta}_k} \\ &+ B_k^w w_k - \Pi_k\delta_k - B_k(H_k e_{k,k} + \sum_{j\in\mathcal{N}_k} H_{kj}\widetilde{C}_j e_{k,j}).\end{aligned}$$

We consider the Lyapunov function $V(\epsilon) = \sum_{k=1}^{N} \underbrace{\epsilon_k^\top X_k^{-1}\epsilon_k}_{V_k(\epsilon_k)}$, where X_k is the solution of (4.58). By completion of squares, the Lie-derivative of V_k can be bounded by

$$\begin{aligned}\dot{V}_k =&\epsilon_k^\top \left(A_k^\top X_k^{-1} + X_k^{-1}A + 2X_k^{-1}B_k H_k\right)\epsilon_k + 2\epsilon_k^\top X_k^{-1}\sum_{j\in\mathcal{N}_k} B_k H_{kj}\widetilde{C}_j\epsilon_j \\ &+ 2\epsilon_k^\top X_k^{-1}\left(B_k^w w_k - \Pi_k\delta_k + \widetilde{\delta}_k - B_k(H_k e_{k,k} + \sum_{j\in\mathcal{N}_k} H_{kj}\widetilde{C}_j e_{k,j})\right) \\ \leq&\epsilon_k^\top\left(A_k^\top X_k^{-1} + X_k^{-1}A - \frac{1}{\lambda^2}X_k^{-1}B_k B_k^\top X_k^{-1}\right)\epsilon_k + 2\epsilon_k^\top X_k^{-1}\sum_{j\in\mathcal{N}_k} B_k H_{kj}\widetilde{C}_j\epsilon_j + \|e_k\|_{W_k}^2 \\ &+ \frac{1}{\mu^2}\epsilon_k^\top X_k^{-1}\left(B_k^w B_k^{w\top} + \Pi_k\Pi_k^\top + B_k B_k^\top\right)X_k^{-1}\epsilon_k + \mu^2\left(\|w_k\|^2 + \|\delta_k\|^2 + \|\widetilde{\delta}_k\|^2\right),\end{aligned}$$

with the weighting matrix W_k defined in (4.60). Next, we apply the Schur-complement to LMI (4.58) and multiply it with X_k^{-1} from both sides, which yields

$$\begin{aligned}X_k^{-1}A_k + A_k^\top X_k^{-1} - \frac{1}{\lambda^2}X_k^{-1}B_k B_k^\top X_k^{-1} + \frac{1}{\mu^2}X_k^{-1}(B_k^w B_k{}^{w}\top + \Pi_k\Pi_k^\top + B_k B_k^\top)X_k^{-1} \\ + X_k^{-1}\widetilde{X}_k X_k^{-1} + ((1+\delta)q_k + p_k)\widetilde{C}_k^\top\widetilde{C}_k + \widetilde{\alpha}_k X_k^{-1} \leq 0.\end{aligned} \tag{4.61}$$

Moreover, the LMI (4.59) can be transformed to

$$\begin{bmatrix} -X_k^{-1}\widetilde{X}_k X_k^{-1} & X_k^{-1}B_k H_{kj_1} - \widetilde{C}_k^\top & \dots & X_k^{-1}B_k H_{kj_{p_k}} - \widetilde{C}_k^\top \\ * & -\delta I_r & & \\ \vdots & & \ddots & \\ * & & & -\delta I_r \end{bmatrix} \preceq 0.$$

With (4.61) and (4.58), we have

$$\dot{V}_k \leq -\epsilon_k^\top \left(((1+\delta)q_k + p_k)\widetilde{C}_k^\top \widetilde{C}_k - \widetilde{\alpha}_k X_k^{-1}\right)\epsilon_k + \sum_{j\in\mathcal{N}_k} \delta\epsilon_j^\top \widetilde{C}_j^\top \widetilde{C}_j \epsilon_j$$
$$+ 2\epsilon_k^\top \sum_{j\in\mathcal{N}_k} \widetilde{C}_k^\top \widetilde{C}_j \epsilon_j + \mu^2\left(\|w_k\|^2 + \|\delta_k\|^2 + \|\widetilde{\delta}_k\|^2\right) + \|e_k\|^2_{W_k}.$$

Summing up $\dot{V}_k$ leads to the Lie-derivative

$$\dot{V} \leq \underbrace{\sum_{k=1}^{N}\left(-(q_k+p_k)\epsilon_k^\top \widetilde{C}_k^\top \widetilde{C}_k\epsilon_k + 2\epsilon_k^\top \widetilde{C}_k^\top \sum_{j\in\mathcal{N}_k}\widetilde{C}_j\epsilon_j\right)}_{\sum_{k=1}^{N}\sum_{j\in j\in\mathcal{N}_k}\|\widetilde{C}_k\epsilon_k - \widetilde{C}_j\epsilon_j\|^2}$$
$$+\sum_{k=1}^{N}\mu^2\left(\|w_k\|^2 + \|\delta_k\|^2 + \|\widetilde{\delta}_k\|^2\right) + \sum_{k=1}^{N}\|e_k\|^2_{W_k} - \widetilde{\alpha}_{min}V, \tag{4.62}$$

with $\widetilde{\alpha}_{min} = \min_k(\widetilde{\alpha}_k) > 0$. Using condition *1)* of the Theorem, we know from (Scardovi & Sepulchre, 2009) that $\zeta_j - \zeta_k \to 0$ exponentially for all $k, j = 1, ..., N$, and therefore, $\|\delta_k\|^2$, $\|\widetilde{\delta}_k\|^2 \to 0$ exponentially.

In the absence of disturbance, $w_k, \eta_k \equiv 0$, we have $e_k \to 0$ exponentially, due to condition 3 of the Theorem and Property 1 of Problem 6. This means that

$$\dot{V} \leq -\sum_{k=1}^{N}\sum_{j\in\mathcal{N}_k}\|\widetilde{C}_k\epsilon_k - \widetilde{C}_j\epsilon_j\|^2 - \alpha_{min}V + \underbrace{\sum_{k=1}^{N}\mu^2\left(\|\delta_k\|^2 + \|\widetilde{\delta}_k\|^2\right) + \sum_{k=1}^{N}\|e_k\|^2_{W_k}}_{\to 0}.$$

Since $-\sum_{k=1}^{N}\sum_{j\in\mathcal{N}_k}\|\widetilde{C}_k\epsilon_k - \widetilde{C}_j\epsilon_j\|^2 \leq 0$, we can immediately conclude $V \to 0$, and therefore $\|z_k - z_j\| \to 0$ for all $k, j = 1, ..., N$, which means that Property 1 of Problem 2 is satisfied.

When the system is affected by disturbances, integrating (4.62) over the interval $[0, T]$ leads to the inequality

$$V(\epsilon(T)) + \int_0^T \sum_{k=1}^{N}\sum_{j\in\mathcal{N}_k}\|\widetilde{C}_k\epsilon_k - \widetilde{C}_j\epsilon_j\|^2 dt \leq \mu^2\sum_{k=1}^{N}\int_0^T \|w_k\|^2 dt$$
$$+\mu^2\sum_{k=1}^{N}\int_0^T \|\delta_k\|^2 + \|\widetilde{\delta}_k\|^2 dt + \sum_{k=1}^{N}\int_0^T \|e_k\|^2_{W_k}dt + \sum_{k=1}^{N}\epsilon_k(0)^\top X_k^{-1}\epsilon_k(0).$$

With the performance property of the distributed observer established in Property 2 of Problem 1, we can conclude

$$\begin{aligned}\int_0^T \sum_{k=1}^N \sum_{j\in\mathcal{N}_k} \|\widetilde{C}_k\epsilon_k - \widetilde{C}_j\epsilon_j\|^2 dt \leq \overline{\kappa}^2 \sum_{k=1}^N \int_0^T \|w_k\|^2 dt + \mu^2 \sum_{k=1}^N \int_0^T \|\delta_k\|^2 + \|\widetilde{\delta}_k\|^2 dt \\ + \overline{\theta}^2 \sum_{k=1}^N \int_0^T \|\eta_k\|^2 dt + I_0 + \sum_{k=1}^N \epsilon_{k,0}^\top X_k^{-1} \epsilon_{k,0},\end{aligned}$$

with $\overline{\kappa} = \sqrt{\mu^2 + \gamma^2 q_{max}}$ and $\overline{\theta} = \gamma$.

In particular, $\widetilde{\delta}_k$ depends on various factors including the control gain matrices H_{kj}, but nevertheless, from the exponential convergence of $\zeta_j - \zeta_k \to 0$ for all $k, j = 1, ..., N$ we can conclude that $\int_0^\infty \|\delta_k\|^2 + \|\widetilde{\delta}_k\|^2 dt$ is finite. Letting $T \to \infty$ shows that (4.51) holds, where

$$\overline{I}_0 = \mu^2 \sum_{k=1}^N \int_0^T \|\delta_k\|^2 + \|\widetilde{\delta}_k\|^2 dt + I_0 + \sum_{k=1}^N \epsilon_{k,0}^\top X_k \epsilon_{k,0}.$$

With Lemma 4.3, Property 2 of Problem 5 is solved. ■

Comparing the control law (4.57) to (4.50), it can be seen that (4.57) includes the neighbours estimates into the control law in the terms of $\sum_{j\in\mathcal{N}_k} H_{kj}(\hat{x}_{k,j} - \Pi_j\zeta_k)$. It makes use of the fact that thanks to the regional estimation scheme (4.41), estimates of the neighbour states $x_j, j \in \mathcal{N}_k$ are available that can be added to the control law in order to improve synchronization performance. An example for this will be shown later. Note these terms are only meant for improving performance and H_{kj} can be set to 0 if some loss of performance is acceptable. Then, (4.57) resembles (4.50).

For the choice of parameters λ and μ, same discussion as done in Remark 4.6 is applicable. Moreover, in order to have better performance guarantees, it is beneficial to solve (4.58) and (4.59) in such a way that W_k is small. Simulations have shown that in fact, it is effective to maximize $\sum_{k=1}^N trace(X_k)$. Smaller W_k allow the subsequent observer design LMIs (4.46) to provide a small performance parameter γ, which improves the closed-loop synchronization performance.

As discussed in Remark 4.4, the observers (4.41) require knowledge about the inputs on the neighboring agents, which may be undesirable. In the case of $H_{kj} = 0, j \in \mathcal{N}_k$, following Corollary presents a method to avoid this drawback.

Corollary 4.1 *Let a group of N agents be given as* (4.30) *with synchronization outputs* (4.2) *and measurement outputs* (4.31)*, and let Problem 5 be solved with the control low*

(4.57) *with* $H_{kj} = 0$ *and the observers* (4.41). *Then, Problem 5 is also solved with the observers*

$$\dot{\hat{x}}_k = \overline{A}_k \hat{x}_k + L_k(y_k - \overline{C}_k \hat{x}_k) + \widetilde{K}_k \left[\hat{x}_{j,j} - \hat{x}_{k,j}\right] j \in \mathcal{N}_k + \overline{B}_k \widetilde{u}_k, \tag{4.63}$$

with

$$\widetilde{u}_k = \begin{bmatrix} \Gamma_k \zeta_k + H_k(\hat{x}_{k,k} - \Pi_k \zeta_k) \\ [\Gamma_j \zeta_k + H_j(\hat{x}_{k,j} - \Pi_j \zeta_k)]_{\mathcal{N}_k} \end{bmatrix}. \tag{4.64}$$

and the consensus gain matrices

$$\widetilde{K}_k = K_k + \begin{bmatrix} 0 \\ [B_j H_j]_{\mathcal{N}_k} \end{bmatrix}.$$

Proof. The difference of the right-hand-side of (4.41) and the right-hand-side of (4.63) can be calculated as

$$\begin{aligned} &\widetilde{K}_k \left[\hat{x}_{j,j} - \hat{x}_{k,j}\right] j \in \mathcal{N}_k + \overline{B}_k \begin{bmatrix} \Gamma_k \zeta_k + H_k(\hat{x}_{k,k} - \Pi_k \zeta_k) \\ [\Gamma_j \zeta_k + H_j(\hat{x}_{k,j} - \Pi_j \zeta_k)]_{\mathcal{N}_k} \end{bmatrix} \\ &- K_k \left[\hat{x}_j^j - \hat{x}_{k,j}\right] j \in \mathcal{N}_k - \overline{B}_k \begin{bmatrix} \Gamma_k \zeta_k + H_k(\hat{x}_{k,k} - \Pi_k \zeta_k) \\ [\Gamma_j \zeta_j + H_j(\hat{x}_{j,j} - \Pi_j \zeta_j)]_{\mathcal{N}_k} \end{bmatrix} \\ &= \overline{B}_k \begin{bmatrix} 0 \\ [\Gamma_j(\zeta_k - \zeta_j) - H_j \Pi_j(\zeta_k - \zeta_j))]_{\mathcal{N}_k} \end{bmatrix}. \end{aligned}$$

As all $\zeta_k - \zeta_j$ converge to 0 exponentially, Problem 5 is solved. ∎

Unlike the distributed observers (4.41), the modified version (4.63) completely removes the inclusion of neighboring states in $\widetilde{u}_k$. The remaining couplings between the observers are only differences between the estimates $\hat{x}_k$, which vanish in the undisturbed case. Moreover, Problem (5) is solved with the same performance parameters κ and θ as in Theorem 4.5 or Theorem 4.6, respectively.

4.2.4 Simulation example

To illustrate the results of this chapter, we consider four agents connected in a topology of a cyclic graph, see Figure 1. The four agents are described by a general third order LTI system

$$\dot{x}_k = \begin{bmatrix} 0 & 1 & 0 \\ 0 & 0 & 1 \\ -a_k & -b_k & -c_k \end{bmatrix} x_k + Bu_k + \begin{bmatrix} 0 \\ 0.1 \\ 0 \end{bmatrix} w_k,$$

with paramters $a_k, b_k, c_k \in \mathbb{R}$. They may, for instance, represent vehicles with position, velocity, and actuator dynamics as the three states. In this example, we assume that each agent takes a relative measurement $y_k = \begin{bmatrix} 1 & 0 & 0 \end{bmatrix} (x_{k+1} - x_k) + 0.5\eta_k$ for $k = 1, 2, 3$ and $y_4 = \begin{bmatrix} 1 & 0 & 0 \end{bmatrix} (x_1 - x_4) + 0.5\eta_4$. Moreover, the outputs to be brought to synchrony are $z_k = \begin{bmatrix} 1 & 0 & 0 \end{bmatrix} x_k$, and the reference trajectories are generated by the systems (4.56) with (S, R) given as

$$S = \begin{bmatrix} 0 & 1 \\ -1 & 0 \end{bmatrix}, \qquad R = \begin{bmatrix} 1 & 0 \end{bmatrix},$$

which corresponds to oscillatory behavour of the output variables.

In the first case, we consider single input agents with $B = \begin{bmatrix} 0 & 0 & 1 \end{bmatrix}^\top$. The Regulator Equations (4.5) are then solved by the matrices

$$\Pi_k = \begin{bmatrix} 1 & 0 \\ 0 & 1 \\ -1 & 0 \end{bmatrix} \quad \Gamma_k = \begin{bmatrix} c_k - a_k & b_k - 1 \end{bmatrix}$$

for all $k = 1, ..., 4$. Moreover, let the system parameters be defined as

$$\begin{aligned} &a_1 = 0.1 && a_2 = 0 && a_3 = 0 && a_4 = -0.1 \\ &b_1 = -0.1 && b_2 = -0.1 && b_3 = -0.1 && b_4 = -0.1 \\ &c_1 = 0 && c_2 = 0 && c_3 = -0.1 && c_4 = -0.1. \end{aligned} \tag{4.65}$$

The four agent matrices $A_1, ..., A_4$ have no common eigenvalue, which is why the observability conditions of Theorem 4.3 is satisfied and we can therefore apply distributed tracking in order to solve Problem 5.

For the distributed regulators (4.57) we choose $\mu = 4, \lambda = 0.1, \delta = 0.8$, and $\widetilde{\alpha}_k = 0.1$ and for the local observers (4.41) we choose $\pi_k = 0.1, \alpha_k = 0.1$ for all k. We achieve guaranteed performance of $\kappa = 7.0$ and $\theta = 5.8$. Interestingly, the same performance is obtained by applying the design method presented in Theorem 4.5, which essentially means that the gains H_{kj} can be set to 0 without loss of performance. Simulations of the synchronizing outputs and the estimation errors are shown in Figure 4.3 and 4.4.

In the second case, let $B = I_3$. The Regulator Equations (4.5) are then solved by the matrices

$$\Pi_k = \begin{bmatrix} 1 & 0 \\ 0 & 1 \\ -1 & 0 \end{bmatrix} \quad \Gamma_k = \begin{bmatrix} 0 & 0 \\ 0 & 0 \\ c_k - a_k & b_k - 1 \end{bmatrix}$$

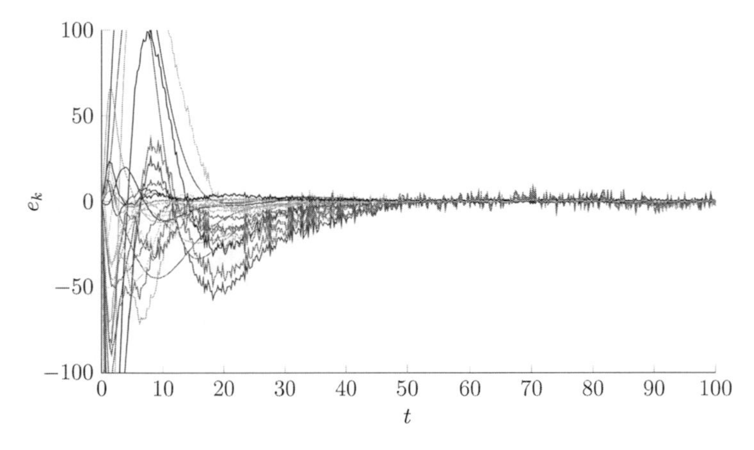

Figure 4.3: Estimation error achieved with measurements disturbed by white noise.

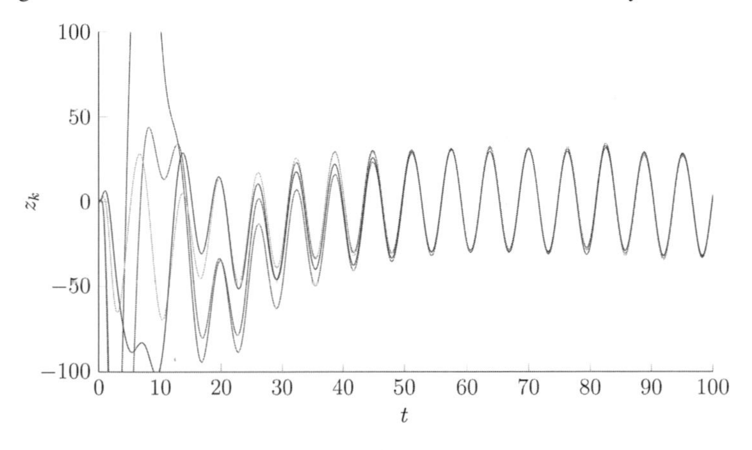

Figure 4.4: Synchronizing behaviour of the outputs z_k.

for all $k = 1, ..., 4$. Moreover, let the system parameters be defined as in (4.65). Here, applying Theorem 4.6, we obtain performance parameters of $\kappa = 5.2$, $\theta = 3.3$ where if we apply Theorem 4.5, performance degrades to $\kappa = 5.9$, $\theta = 4.3$.

Remark 4.7 *If e.g. the agent parameters are*

$$
\begin{aligned}
&a_1 = 0 && a_2 = 0 && a_3 = 0 && a_4 = 0 \\
&b_1 = -0.1 && b_2 = -0.1 && b_3 = -0.1 && b_4 = -0.1
\end{aligned}
$$

$$c_1 = 0 \qquad c_2 = 0 \qquad c_3 = -0.1 \qquad c_4 = -0.1,$$

then the observability conditions of Theorem 4.3 are not satisfied. Hence, in order to solve Problem 4, Theorem 4.6 cannot be applied, but instead, Theorem 4.1 can be applied.

4.3 Extensions of Regional Estimation-based Control

Now, we present some extensions of the results presented in the previous Section, where the extensions aim at expanding the system class and the control goal, and reducing model information requirements. In all of these extensions, the regional estimation scheme plays a crucial role at enabling the respective control laws.

4.3.1 Physically interconnected agents

Regarding the synchronization problems for MAS, most results in the literature deal with autonomous agents, i.e. subsystems with no physical coupling. Thus, all couplings are introduced by controller design

We consider a group of N LTI agents described by the differential equation

$$\dot{x}_k = A_k x_k + \sum_{j \in \mathcal{N}_k} A_{kj} x_j + B_k u_k + B_k^w w_k, \tag{4.66}$$

where for all $k \in \mathcal{N}$, $x_k(t) \in \mathbb{R}^{n_k}$ is the state of the agent, $u_k(t) \in \mathbb{R}^{m_k}$ is the respective control input, and w with $w_k(t) \in \mathbb{R}^{v_k}$ is the $\mathcal{L}_2$-integrable disturbance affecting agent k. The synchronization output z_k and the measurements output y_k are defined in (4.2) and (4.31), respectively. As it can be seen in (4.66), the agents' states are interconnected in the sense that each agent's state is affected by its neighbours' states. As usual, the neighbourhood is defined by the directed graph $\mathcal{G} = (\mathcal{V}, \mathcal{E})$.

To be precise, we have three separate graphs representing the three different layers of interaction

1. Dynamic interaction in (4.66), represented by $\mathcal{G}_{dyn} = (\mathcal{V}, \mathcal{E}_{dyn})$, where $(j,k) \in \mathcal{E}_{dyn}$ if $A_{kj} \neq 0$.

2. Measurement coupling in (4.31), represented by $\mathcal{G}_{meas} = (\mathcal{V}, \mathcal{E}_{meas})$, where $(j,k) \in \mathcal{E}_{meas}$ if $C_{kj} \neq 0$.

3. Communication channels, represented by $\mathcal{G}_{comm} = (\mathcal{V}, \mathcal{E}_{comm})$, where $(j,k) \in \mathcal{E}_{comm}$ if agent j can send information to agent k.

The unified graph $\mathcal{G}$ is subsequently defined as $\mathcal{E} = \mathcal{E}_{comm} \supset \mathcal{E}_{dyn}, \mathcal{E}_{meas}$. This definition is in alignment to Assumption 4.1 and its discussion in Remark 4.2. Moreover, we make the following assumption:

Assumption 4.5 *For all $k \in \mathcal{N}$, the vertices of the inclusive neighbourhood $\{v_j | j \in \{k \cup j \in \mathcal{N}_k\}\}$ form an independent component of the interconnection graph $\mathcal{G}_{dyn}$.*

This is a technical assumption, which is made for the sake of simplicity. Our recent publication (Wu, Ugrinovskii & Allgöwer, 2015) deals with the case when it is not satisfied. If $\mathcal{G}$ can be designed, this assumption presents a clear requirement where additional edges have to be added. However, if $\mathcal{G}_{dyn}$ is strongly connected, Assumption 4.5 will require that $\mathcal{G}$ is an all-to-all graph, resulting in the global estimation scheme discussed in Section 4.2.1.

For the class of interconnected agents (4.66), the Regulator Equations need to be modified to

$$\begin{aligned} A_k\Pi_k + \sum_{j\in\mathcal{N}_k} A_{kj}\Pi_j + B_k\Gamma_k &= \Pi_k S \\ \widetilde{C}_k\Pi_k &= R, \end{aligned} \tag{4.67}$$

where we will again make Assumption (4.27) as exponential divergence of the trajectories is not desirable.

The definitions of the local estimates $\hat{x}_k$ (4.38), combined states $\overline{x}_k$, and estimation error e_k (4.39) are kept for the interconnected agents. With the distributed observers (4.41) and the agents (4.66), it holds for the estimation error e_k that

$$\dot{e}_k = (\overline{A}_k - L_k\overline{C}_k)e_k + K_k\left[e_{j,j} - e_{k,j}\right]_{j\in\mathcal{N}_k} - L_k\eta_k + B_k^w w_k + \Delta_k, \tag{4.68}$$

with

$$\Delta_k = \begin{bmatrix} 0_{n_k\times n_k} \\ \left[-\sum_{i\in\mathcal{N}_j\setminus(\mathcal{N}_k\cup k)} A_{ji}x_i\right]_{j\in\mathcal{N}_k} \end{bmatrix}, \quad \overline{B}_k^w = \operatorname{diag}\left[E_j\right]_{j\in\mathcal{N}_k}, \quad \overline{w}_k = \left[w_j\right]_{j\in\mathcal{N}_k}.$$

Thus, for all $k \in \mathcal{N}$ the error differential equations are composed as follows:

$$\begin{array}{lll} (\overline{A}_k - L_k\overline{C}_k)e_k & \ldots & \text{Classical Luenberger observer error differential equation} \\ +K_k\left[e_{j,j} - e_{k,j}\right]_{j\in\mathcal{N}_k} & \ldots & \text{Consensus term} \\ -L_k\eta_k + \operatorname{diag}\left[E_j\right]_N k\left[w_j\right]_N k & \ldots & \text{exogeneous and measurement disturbance} \\ +\Delta_k & \ldots & \text{additional interconnections of the neighbours} \end{array} \tag{4.69}$$

Note that if Assumption 4.5 is satisfied, it holds that $\Delta_k \equiv 0$ for all $k \in \mathcal{N}$.

In this section, we propose an extension of the synchronization algorithm, where we use coupled LMIs similar to the observer design (4.46). Thereby, we can both address the case of interconnected agents and optimize the performance guarantees.

The controller is now proposed as

$$\dot{\zeta}_k = S\zeta_k + \underbrace{\sum_{j\in\mathcal{N}_k}(\zeta_j - \zeta_k)}_{\delta_k} \tag{4.70}$$

$$u_k = \Gamma_k\zeta_k + H_k(\hat{x}_{k,k} - \Pi_k\zeta_k) + \sum_{j\in\mathcal{N}_k} H_{kj}(\hat{x}_{k,j} - \Pi_j\zeta_k), \tag{4.71}$$

where the matrices $S \in \mathbb{R}^{\nu\times\nu}, R \in \mathbb{R}^{\tilde{r}\times\nu}, \Pi_k \in \mathbb{R}^{n_k\times\nu}$ and $\Gamma_k \in \mathbb{R}^{m_k\times\nu}$ are obtained from the Francis equations for interconnected agents (4.67).

With these definitions, a solution to Problem 5 is given in the following Theorem.

Theorem 4.7 *Consider a group of N interconnected LTI systems* (4.30), *with the communication topology represented by a connected graph $\mathcal{G} = \{\mathcal{V}, \mathcal{E}\}$ and the control inputs given as* (4.71) *with local reference generators* (4.70) *and observers* (4.41). *Suppose the following conditions hold:*

1. *There exists a positiv integer ν and matrices $S \in \mathbb{R}^{\nu\times\nu}$, $R \in \mathbb{R}^{\tilde{r}\times\nu}$, $\Pi_k \in \mathbb{R}^{n_k\times\nu}$, $\Gamma_k \in \mathbb{R}^{m_k\times\nu}$ such that (4.5),(4.27) are satisfied for all $k \in \mathcal{N}$.*

2. *Let $\lambda, \mu > 0$. For all $k \in \mathcal{N}$, let $\mathcal{N}_k = \{j_1, ..., j_{p_k}\}$ and let the collection of matrices $X_k \succ 0$, $\widetilde{X}_k \succ 0$, $H_{kj_1}, ..., H_{kj_{p_k}}$ and parameters π_k be a solution of the two LMIs*

$$\begin{bmatrix} Q_k + \widetilde{X}_k & X_k \\ X_k & -\left((q_k+p_k)\widetilde{C}_k^\top\widetilde{C}_k + \pi_k q_k I_{n_k}\right)^{-1}\end{bmatrix} \preceq 0 \tag{4.72}$$

$$\begin{bmatrix} -\widetilde{X}_k & (A_{kj_1}+B_kH_{kj_1}) - X_k\widetilde{C}_k^\top\widetilde{C}_{j_1} & \dots & (A_{kj_{p_k}}+B_kH_{kj_{p_k}}) - X_k\widetilde{C}_k^\top\widetilde{C}_{j_{p_k}} \\ * & -\pi_{j_1}I_{n_{j_1}} & & \\ \vdots & & \ddots & \\ * & & & -\pi_{j_{p_k}}I_{n_{j_{p_k}}}\end{bmatrix} \preceq 0, \tag{4.73}$$

with

$$Q_k = A_kX_k + X_kA_k^\top - \frac{1}{\lambda^2}B_kB_k^\top + \frac{1}{\mu^2}(B_k^wB_k^{w\top} + \Pi_k\Pi_k^\top + I_{n_k}) + \widetilde{\alpha}_kX_k \tag{4.74}$$

and arbitrarily small $\widetilde{\alpha}_k > 0$.

3. *The observers* (4.41) *are designed such that they solve Problem 6 with the weighting matrices*

$$W_k = \begin{bmatrix} \frac{1}{\lambda} X_k^{-1\top} B_k \\ \lambda H_{kj_1}^\top \\ \vdots \\ \lambda H_{kj_{p_k}}^\top \end{bmatrix} \begin{bmatrix} \frac{1}{\lambda} X_k^{-1\top} B_k \\ \lambda H_{kj_1}^\top \\ \vdots \\ \lambda H_{kj_{p_k}}^\top \end{bmatrix}^\top, \tag{4.75}$$

and some performance paramter $\gamma > 0$.

Then the control laws (4.71) *with the matrices* H_k *defined as*

$$H_k = -\frac{1}{\lambda^2} B_k^\top X_k^{-1} \tag{4.76}$$

and the solution matrices H_{kj} *solve Problem 5 with any* $\kappa > \sqrt{\mu^2 + \gamma^2}$ *and* $\theta > \gamma$.

Proof. The proof is shown in Appendix B.2. ■

4.3.2 Unknown neighbour models

An additional assumption that was implicitly made in the observer setup (4.41) is that every agent also knows the model matrices (A_j, B_j, E_j) of its neighbours. This is not necessarily restrictive as the neighbouring agents may communicate these matrices and obviously, every agent is assumed to know its own model. However, in order to achieve improved adaptability of the global system, it can be desirable to change the observers (4.41) in such a way that the exact matrices (A_j, B_j, E_j) of the neighbours are not needed.

As in Section 4.2, we consider autonomous agents (4.30) with the synchronization outputs (4.2), and the synchronizing control law (4.50). Concerning the measurements, we consider the special case of (4.31), where

$$y_k = \left([z_j]_{j \in \mathcal{N}_k} - \mathbf{1}_{p_k} \otimes z_k \right) + \eta_k, \tag{4.77}$$

i.e. the output $y_k(t) \in \mathbb{R}^{p_k r}$ that each agent obtains, consists of the relative outputs $z_j - z_k, j \in \mathcal{N}_k$ with additional disturbance. The interconnection topology is given by the directed, connected graph $\mathcal{G} = (\mathcal{V}, \mathcal{E})$ and assumption 4.1 is again used here.

The following results are based on the approach that instead of estimating the neighbours' state $x_j, j \in \mathcal{N}_k$, projected states are estimated. From the perspective of an agent k, suppose that all neighbours $j \in \mathcal{N}_k$ are perfectly initialized and regulated in the sense that the regulation error $\epsilon_j \equiv 0, j \in \mathcal{N}_k$, where we assume for now that $\delta_k = 0, k \in \mathcal{N}$ in (4.49). Then,

$$x_j = \Pi_j \zeta_j$$

for all $j \in \mathcal{N}_k$. Thus, for the relative output $z_j - z_k$ it holds that

$$z_j - z_k = \widetilde{C}_j x_j - \widetilde{C}_k x_k = R\zeta_j - \widetilde{C}_k x_k.$$

With respect to the regional estimation scheme presented in 4.2.1, here, it seems unnecessary to estimate the neighbours' state x_j. Instead, ζ_j could be used and ϵ_k merely acts as disturbance. If $\epsilon_k \to 0$, then so would the estimation error e_k. However, it has to be taken into account that the differential equation of ϵ_k, as shown in Section 4.2, are affected by e_k. This discussion gives an intuition about the solution to Problem 5, which is presented next.

For the solution to (4.5), we make the following assumption:

Assumption 4.6 *Let the pair* (S, R) *satisfying* (4.27) *be given and let the matrices* $\Pi_k \in \mathbb{R}^{n_k \times \nu}$ *and* $\Gamma_k \in \mathbb{R}^{m_k \times \nu}$ *be a solution to the Regulator equations* (4.5). *Then, we assume that every* Π_k *has full column rank. Moreover, let* Π_k^{-1} *be defined as the left-inverse in the sense that* $\Pi_k^{-1}\Pi_k = I_\nu$, *then we assume that* $C_k = R\Pi_k^{-1}$.

This assumption in fact means that every state of the internal model is projected to a state of the agents x_k and is a mild assumption if $\nu \leq n_k$ for all $k \in \mathcal{N}$. One example will be given later.

Let the vector of local estimates be $\hat{x}_k = \begin{bmatrix} \hat{x}_{k,k}^\top & [\hat{x}_{k,j}^\top]_{j\in\mathcal{N}_k} \end{bmatrix}^\top \in \mathbb{R}^{\sigma_k}$ with $\sigma_k = n_k + p_k\nu$ and the corresponding error vector

$$e_k = \begin{bmatrix} e_{k,k} \\ [e_{k,j}]_{j\in\mathcal{N}_k} \end{bmatrix} = \begin{bmatrix} x_k - \hat{x}_{k,k} \\ [\Pi_j^{-1}x_j - \hat{x}_{k,j}]_{j\in\mathcal{N}_k} \end{bmatrix},$$

i.e. the agents estimate their neighbours' projected states $\Pi_j^{-1}x_j \in \mathbb{R}^n$ in contrast to Section 4.2.1, where the agents estimate their neighbours' full states $x_j \in \mathbb{R}^{n_j}$.

The observer differential equations are now proposed as

$$\begin{aligned} \dot{\hat{x}}_k =& \overline{A}_k\hat{x}_k + L_k(y_k - \overline{C}_k\hat{x}_k) + \begin{bmatrix} B_k u_k \\ [\Pi_j^{-1}B_jH_j\hat{\epsilon}_j]_{j\in\mathcal{N}_k} \end{bmatrix} \\ &+ K_k\left[\Pi_j^{-1}\hat{x}_{j,j} - \hat{x}_{k,j}\right]_{j\in\mathcal{N}_k} + \widetilde{K}_k[\Delta_j\hat{\epsilon}_j]_{j\in\mathcal{N}_k} \end{aligned} \tag{4.78}$$

with the definitions

$$\begin{aligned} \overline{A}_k &= \begin{bmatrix} A_k & 0 \\ 0 & I_{p_k} \otimes S \end{bmatrix}, & \overline{C}_k &= \begin{bmatrix} -\mathbf{1}_{p_k} \otimes C_k & I_{p_k} \otimes R \end{bmatrix} \\ \hat{\epsilon}_j &= \hat{x}_{j,j} - \Pi_j \zeta_j, & \Delta_j &= \Pi_j^{-1} A_j - S \Pi_j^{-1}, \end{aligned} \tag{4.79}$$

and the initial condition $\hat{x}_k(0) = 0$. The filter gains to be designed are $L_k \in \mathbb{R}^{\sigma_k \times p_k r}$ and $K_k, \widetilde{K}_k \in \mathbb{R}^{\sigma_k \times p_k n}$. The matrix Δ_j in (4.79) can be interpreted as the *projection error matrix,* i.e. the amount by which the projected agent state feedback matrix differs from the virtual exosystem matrix S.

Remark 4.8 *As it can be seen in* (4.78)*, each agent j broadcasts three variables to its out-neighbourhood:*

- *projected state estimate* $\Pi_j^{-1} \hat{x}_{j,j}$
- $\Delta_j \hat{\epsilon}_j$
- $\Pi_j^{-1} B_j H_j \hat{\epsilon}_j$.

This is in contrast to (4.41)*, where only state estimates $\hat{x}_{j,j}$ were transmitted. The additional transmission is required here because the agents do not have the model of their neighbours.*

In order to establish the conditions for filter design, we need to calculate the closed-loop error differential equation, which are composed out of the regulation error and the estimation error. For the regulation error, we have the derivative

$$\begin{aligned} \dot{\epsilon}_k =& \dot{x}_k - \Pi_k \dot{\zeta}_k \\ =& A_k x_k + B_k \Gamma_k \zeta_k + B_k H_k (\hat{x}_{k,k} - \Pi_k \zeta_k) + B_k^w w_k - \Pi_k S \zeta_k - \Pi_k \underbrace{\sum_{j \in \mathcal{N}_k} (\zeta_j - \zeta_k)}_{\delta_k} \end{aligned}$$

and moreover, it follows from (4.5) that $B_k \Gamma_k \zeta_k - \Pi_k S \zeta_k = -A_k \Pi_k \zeta_k$ and therefore

$$\begin{aligned} \dot{\epsilon}_k &= A_k x_k - A_k \Pi_k \zeta_k + B_k H_k (x_k - \Pi_k \zeta_k) + B_k^w w_k - \Pi_k \delta_k - B_k H_k (x_k - \hat{x}_{k,k}) \\ &= (A_k + B_k H_k) \epsilon_k + B_k^w w_k - \Pi_k \delta_k - B_k H_k e_{k,k}. \end{aligned}$$

The estimation errors are more difficult to calculate: We have the derivative

$$\begin{aligned}\dot{e}_k = &\begin{bmatrix} A_k x_k + B_k u_k + B_k^w w_k - A_k \hat{x}_{k,k} - B_k u_k \\ [\underbrace{\Pi_j^{-1} A_j x_j + \Pi_j^{-1} B_j \Gamma_j \zeta_j - S\hat{x}_{k,j}}_{\mathbf{I}} + \Pi_j^{-1} E_j w_j]_{j\in\mathcal{N}_k} \end{bmatrix} \\ &- L_k \Big[\underbrace{C_j x_j - C_k x_k - R\hat{x}_{k,j} + C_k \hat{x}_{k,k}}_{\mathbf{II}}\Big]_{j\in\mathcal{N}_k} \\ &- L_k \eta_k - K_k \Big[\underbrace{\Pi_j^{-1}\hat{x}_{j,j} - \hat{x}_{k,j}}_{\mathbf{III}}\Big]_{j\in\mathcal{N}_k} - \widetilde{K}_k \Big[\underbrace{\Delta_j \hat{\epsilon}_j}_{\mathbf{IV}}\Big]_{j\in\mathcal{N}_k},\end{aligned} \tag{4.80}$$

where for the various terms in (4.80), we can establish the equations

$$\begin{aligned}&\mathbf{I}: \\ &\quad \Pi_j^{-1} A_j x_j + \Pi_j^{-1} B_j \Gamma_j \zeta_j - S\hat{x}_{k,j} \\ &\quad = \Pi_j^{-1} A_j \epsilon_j + \Pi_j^{-1} A_j \Pi_j (\zeta_j - \hat{x}_{k,j}) + \Pi_j^{-1} B_j \Gamma_j (\zeta_j - \hat{x}_{k,j}) \\ &\quad = \Pi_j^{-1} A_j \epsilon_j + S e_{k,j} - S\Pi_j^{-1}\epsilon_j = S e_{k,j} + (\Pi_j^{-1} A_j - S\Pi_j^{-1})\epsilon_j = S e_{k,j} + \Delta_j \epsilon_j \\ &\mathbf{II}: \\ &\quad C_j x_j - C_k x_k - R\hat{x}_{k,j} + C_k \hat{x}_{k,k} = (C_j - R\Pi_j^{-1})\epsilon_j - C_k e_{k,k} + R e_{k,j} \\ &\quad = R e_{k,j} - C_k e_{k,k} \\ &\mathbf{III}: \\ &\quad \Pi_j^{-1}\hat{x}_{j,j} - \hat{x}_{k,j} = \Pi_j^{-1}(x_j - e_{j,j}) - \hat{x}_{k,j} = e_{k,j} - \Pi_j^{-1} e_{j,j} \\ &\mathbf{IV}: \\ &\quad \Delta_j \hat{\epsilon}_j = \Delta_j (\hat{x}_{j,j} - \Pi_j \zeta_j) = \Delta_j \epsilon_j - \Delta_j e_{j,j}.\end{aligned}$$

Thus, we summarize these terms to

$$\begin{aligned}\dot{e}^k = &\begin{bmatrix} A_k e_{k,k} + B_k^w w_k \\ [S e_{k,j} + \Pi_j^{-1} E_j w_j]_{j\in\mathcal{N}_k} \end{bmatrix} - L^k \eta_k + L^k \Big[C_k e_{k,k} - R e_{k,j}\Big]_{j\in\mathcal{N}_k} \\ &- K^k \Big[e_{k,j} - \Pi_j^{-1} e_{j,j}\Big]_{j\in\mathcal{N}_k} + \begin{bmatrix} 0 \\ [\Delta_j \epsilon_j]_{j\in\mathcal{N}_k} \end{bmatrix} - \widetilde{K}^k \Big[\Delta_j \epsilon_j - \Delta_j e_{j,j}\Big]_{j\in\mathcal{N}_k} \\ = &\left(A^k - L^k C^k - K^k [0_{p_k\nu\times n_k} I_{p_k\nu}]\right) e^k + \widetilde{K}^k \Big[\Delta_j e_{j,j}\Big]_{j\in\mathcal{N}_k} + K^k [\Pi_j^{-1} e_{j,j}]_{j\in\mathcal{N}_k} \\ &+ \left(\begin{bmatrix} 0 \\ I_{p_k v} \end{bmatrix} - \widetilde{K}^k\right) [\Delta_j \epsilon_j]_{j\in\mathcal{N}_k} + \underbrace{\begin{bmatrix} B_k^w & 0 \\ 0 & I_{p_k\nu} \end{bmatrix}}_{\overline{B}_k^w} \underbrace{\begin{bmatrix} w_k \\ [\Pi_j^{-1} E_j w_j]_{j\in\mathcal{N}_k} \end{bmatrix}}_{\overline{w}_k} - L^k \eta_k.\end{aligned} \tag{4.81}$$

The estimation error differential equation (4.81) differs significantly from the estimation error in Section 4.2.1. In particular, the role of the *projection error matrix* should be emphasized, which is a gain for both the effect of the neighbours estimation error $e_{j,j}$ and the neighbours regulation error ϵ_j on e_k. Here the role of the additional observer gain matrix $\widetilde{K}_k$ is to limit the influence of the *projection error*, such that closed-loop stability can be ensured.

The process of designing the observer and regulation gains has to take account both on the effect of regulation error on estimation error and vice versa. For this purpose, two Lemmas are introduced in the following in order to cover these two issues.

First, we ensure certain $\mathcal{H}_\infty$-performance of the regulation error with respect to the estimation error and the disturbances, which is given in the following Lemma:

Lemma 4.4 *Consider an agent* (4.30) *with the control law* (4.50), *and suppose* $e_{k,k} = x_k - \hat{x}_{k,k}$ *is an exogenous* $\mathcal{L}_2$ *integrable disturbance. Let there be performance parameters* $\mu, \lambda > 0$, *such that for each* $k = k \in \mathcal{N}$, *the algebraic Riccati equation*

$$X_k A_k + A_k^\top X_k + \Delta_k^\top \Delta_k + \hat{W}_k - X_k \left(\frac{1}{\lambda^2} B_k B_k^\top - \frac{1}{\mu^2} B_k^w B_k^{w\top} - \frac{1}{\mu^2} \Pi_k \Pi_k^\top \right) X_k = 0 \tag{4.82}$$

has a positive definite symmetric solution X_k *such that* $A_k - \frac{1}{\lambda^2} B_k B_k^\top X_k$ *is a Hurwitz matrix. Then, with the control gain*

$$H_k = -\frac{1}{\lambda^2} B_k^\top X_k, \tag{4.83}$$

the following $\mathcal{H}_\infty$*-type condition holds*

$$\int_0^\infty \|\epsilon_k\|^2_{\Delta_k^\top \Delta_k + \hat{W}_k} dt \leq \mu^2 \int_0^\infty \|w_k\|^2 dt + \mu^2 \int_0^\infty \|\delta_k\|^2 dt + \lambda^2 \int_0^\infty \|H_k e_{k,k}\|^2 + \hat{I}_{0,k} \tag{4.84}$$

with a constant $\hat{I}_{0,k}$ *dependent on the initial conditions of* ϵ_k *and* ζ_k.

Proof. The proof is can be derived in analogy to the first part of the proof of Theorem 4.5, which can be found in Appendix B.1. ■

Next, we establish conditions for the observer design in order to ensure certain $\mathcal{H}_\infty$-performance of the estimation error with respect to the regulation error and disturbances.

We define the matrices

$$\begin{aligned} Q_k =& P_k\overline{A}_k + \overline{A}_k^\top P_k - G_k^\top \overline{C}_k - (G_k\overline{C}_k)^\top + q_k \begin{bmatrix} \Pi_k^{-1\top}\widetilde{P}_k\Pi_k^{-1} + \Delta_k^\top \widetilde{P}_k\Delta_k & 0 \\ 0 & 0 \end{bmatrix} \\ &- F_k\left[0_{p_k\nu\times n_k} I_{p_k\nu}\right] - \left(F_k\left[0_{p_k\nu\times n_k} I_{p_k\nu}\right]\right)^\top + \alpha_k P_k, \\ W_k =& \begin{bmatrix} \frac{1}{\lambda^2}X_kB_kB_k^\top X_k & 0 \\ 0 & 0 \end{bmatrix}, \end{aligned} \tag{4.85}$$

where $P_k \in \mathbb{R}^{\sigma_k\times\sigma_k}$ and $\widetilde{P}_k \in \mathbb{R}^{v\times v}$ are symmetric, positive definite matrices for all $k = 1, ..., N$. Moreover, introduce matrices $F_k, \widetilde{F}_k \in \mathbb{R}^{\sigma_k\times p_k\nu}$, $G_k \in \mathbb{R}^{\sigma_k\times p_k r}$, and an arbitrarily small $\alpha_k > 0$ which will later be used as a design parameter. With these definitions, we are ready to present the design procedure for the observers.

Lemma 4.5 *Consider a group of N agents described by* (4.30), (4.31) *and suppose a collection of linear matrix inequalities*

$$\left[\begin{array}{ccc:ccc} Q_k + W_k & P_k\overline{B}_k^w & -G_k & F^k & \widetilde{F}^k & P^k\begin{bmatrix} 0 \\ I_{p_k\nu}\end{bmatrix} - \widetilde{F}^k \\ * & -\beta I & 0 & 0 & 0 & 0 \\ * & 0 & -\beta I & 0 & 0 & 0 \\ \hdashline * & 0 & 0 & -\mathrm{diag}[\widetilde{P}^j]_{j\in\mathcal{N}_k} & 0 & \\ * & 0 & 0 & 0 & -\mathrm{diag}[\widetilde{P}^j]_{j\in\mathcal{N}_k} & 0 \\ * & 0 & 0 & 0 & 0 & -\frac{1}{q_{max}}I_{p_k v} \end{array}\right] \preceq 0 \tag{4.86}$$

has a feasible solution in the matrix variables $P^k > 0, F^k, \widetilde{F}^k, G^k$, for all $k \in \mathcal{N}$ and a scalar variable $\beta > 0$. Then, the observers (4.78) with the parameters

$$L_k = (P_k)^{-1}G_k, \quad K_k = (P_k)^{-1}F_k, \quad \widetilde{K}_k = (P_k)^{-1}\widetilde{F}_k, \tag{4.87}$$

satisfy $\mathcal{H}_\infty$-type performance of the estimation error in the sense that

$$\begin{aligned} \lambda^2\sum_{k=1}^N\int_0^\infty \|H_ke_{k,k}\|^2dt \leq& \beta\sum_{k=1}^N(\widetilde{q}_k + 1)\int_0^\infty \|w_k\|^2dt + \beta\sum_{k=1}^N\int_0^\infty \|\eta_k\|^2dt \\ &+ \sum_{k=1}^N\int_0^\infty \|\epsilon_k\|^2_{\Delta_k^\top\Delta_k}dt + I_0 \end{aligned} \tag{4.88}$$

where ϵ_k is regarded to be an exogenous $\mathcal{L}_2$ integrable disturbance and $\widetilde{q}_k = \overline{\sigma}(\Pi_k^{-1}B_k^w)^2q_k$.

Proof. The proof is shown in Appendix B.3 ■

Note that the LMIs (4.86) involve parameters of the model of agent k only, and do not involve parameters of its neighbours' models, which was the main motivation in this section. Moreover, while solving the LMI (4.86), the parameter β can be minimized.

Now, we are ready to present our main result concerned with stability and performance of the proposed observer-based synchronization algorithm.

Theorem 4.8 *Consider a group of N LTI systems* (4.30) *with the control laws* (4.49),(4.50) *and the distributed observers* (4.78), *which communicate with a topology represented by the directed connected graph* $\mathcal{G} = \{\mathcal{V}, \mathcal{E}\}$. *Suppose the following conditions hold:*

1. *The regulator equations* (4.5) *have solutions with a common observable pair* (S, R), *with* $\sigma(S) \subset j\mathbb{R}$.
2. *The control gain matrices* H_k *are defined as in* (4.83) *where* X_k *is a solution to the Riccati equation* (4.82), *with* $\hat{W}_k = 2(q_k + p_k)\widetilde{C}_k^\top \widetilde{C}_k$, *such that* $A_k - \frac{1}{\lambda^2} B_k B_k^\top X_k$ *is a Hurwitz matrix.*
3. *The observer gain matrices* L_k, K_k, *and* $\widetilde{K}_k$ *are defined as in* (4.87), *where* $P_k, F_k, \widetilde{F}_k, G_k, \beta$ *solve* (4.86).

Then, the proposed observer-based control law solves Problem 5 with
$\kappa > \sqrt{\mu^2 + \beta(\widetilde{q}_{max} + 1)}$, *where* $\widetilde{q}_{max} = \max_k(\overline{\sigma}(\Pi_k^{-1} B_k^w)^2 q_k)$, *and* $\theta > \sqrt{\beta}$.

Proof.

First, we show that Property 2 of Problem 5 holds. For this purpose, we consider the sum of (4.84) over all $k \in \mathcal{N}$, which is

$$\sum_{k=1}^{N} \int_0^\infty \|\epsilon_k\|^2_{\Delta_k^\top \Delta_k + \hat{W}_k} dt \leq \mu^2 \sum_{k=1}^{N} \int_0^\infty (\|w_k\|^2 + \|\delta_k\|^2)\, dt + \lambda^2 \sum_{k=1}^{N} \int_0^\infty \|H_k e_{k,k}\|^2 + \sum_{k=1}^{N} \hat{I}_{0,k}.$$

For the term $\int_0^\infty \|H_k e_{k,k}\|^2$ we can apply the inequality (4.88), which gives us the inequalities

$$\sum_{k=1}^{N} \int_0^\infty \|\epsilon_k\|^2_{\Delta_k^\top \Delta_k + \hat{W}_k} dt \leq \mu^2 \sum_{k=1}^{N} \int_0^\infty (\|w_k\|^2 + \|\delta_k\|^2)\, dt + \sum_{k=1}^{N} \hat{I}_{0,k}$$
$$+ \beta \sum_{k=1}^{N} (\widetilde{q}_k + 1) \int_0^\infty \|w_k\|^2 dt + \beta \sum_{k=1}^{N} \int_0^\infty \|\eta_k\|^2 dt + \sum_{k=1}^{N} \int_0^\infty \|\epsilon_k\|^2_{\Delta_k^\top \Delta_k} dt + I_0$$

$$\sum_{k=1}^{N}\int_0^\infty \|\epsilon_k\|^2_{\hat{W}_k}dt \leq \sum_{k=1}^{N}\int_0^\infty \left(\overline{\kappa}^2\|w_k\|^2+\beta\|\eta_k\|^2\right)dt + \underbrace{\sum_{k=1}^{N}\int_0^\infty \mu^2\|\delta_k\|^2 dt + \sum_{k=1}^{N}\hat{I}_{0,k} + I_0}_{\widetilde{I}_0},$$

with $\overline{\kappa} = \sqrt{\mu^2+\beta(\widetilde{q}_{max}+1)}$. The left-hand-side can be bounded by the calculations shown in (B.3) and with Lemma 4.3, the performance criterion (4.33) holds with $\kappa > \overline{\kappa}$ and $\theta > \sqrt{\beta}$.

In order to show Property 1 of Problem 5, we consider the case of $w_k, \eta_k \equiv 0$, where we have that

$$\sum_{k=1}^{N}\int_0^\infty \|\epsilon_k\|^2_{\hat{W}_k}dt \leq \widetilde{I}_0.$$

The trajectory $\epsilon_k(t)$ is clearly uniformly continuous. Therefore, with Barbalat's Lemma, we can conclude that $\|\epsilon_k\|^2_{\hat{W}_k} \to 0$, which shows Property 1 of Problem 5. ■

Simulation Example:

We illustrate our method with four third order LTI agents

$$\dot{x}_k = \begin{bmatrix} 0 & 1 & 0 \\ 0 & 0 & 1 \\ -a_k & -b_k & -c_k \end{bmatrix} x_k + \begin{bmatrix} 0 \\ 0 \\ 1 \end{bmatrix} u_k + \begin{bmatrix} 0 \\ 0.5 \\ 0 \end{bmatrix} w_k,$$

with parameters $a_k, b_k, c_k \in \mathbb{R}$ and $y_k = \begin{bmatrix} 1 & 0 & 0 \end{bmatrix}(x_{k+1}-x_k) + 0.1\eta_k$ for $k = 1,2,3$ and $y_4 = \begin{bmatrix} 1 & 0 & 0 \end{bmatrix}(x_1 - x_4) + 0.1\eta_4$. The system parameters are defined same as before as

$$\begin{array}{llll} a_1 = 0.1 & a_2 = 0 & a_3 = 0 & a_4 = -0.1 \\ b_1 = -0.1 & b_2 = -0.1 & b_3 = -0.1 & b_4 = -0.1 \\ c_1 = 0 & c_2 = 0 & c_3 = -0.1 & c_4 = -0.1 \end{array}$$

and the reference trajectories are generated with (S, R) given as

$$S = \begin{bmatrix} 0 & 1 \\ -1 & 0 \end{bmatrix}, \qquad R = \begin{bmatrix} 1 & 0 \end{bmatrix},$$

which corresponds to oscillatory behaviour of the output variables. Thus, the solution of (4.5) are the matrices

$$\Pi_k = \begin{bmatrix} 1 & 0 \\ 0 & 1 \\ -1 & 0 \end{bmatrix} \quad \Gamma_k = \begin{bmatrix} c_k - a_k & b_k - 1 \end{bmatrix}$$

for all $k = 1, ..., 4$.

The four agents are connected with a cyclic-type, directed communication topology. The observers and controllers were designed using the method presented in Theorem 1. There, for the Ricatti equation (4.82) we chose $\mu = 4$, and $\lambda = 0.1$. Figures 4.5 and 4.6 show the simulations results for the obtained controller. Between $t = 50$ and $t = 60$, an input disturbance w_k is applied, which affects both the estimation and synchronization error, but is efficiently attenuated afterwards with the guaranteed attenuation gain of $\kappa = 16.6$. This κ is rather high, which reflects the lack of knowledge on the neighbouring agents' models.

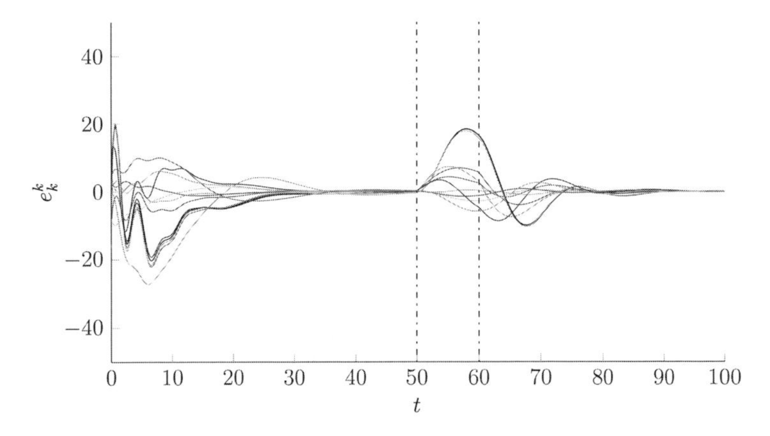

Figure 4.5: Estimation error.

4.3.3 Regional estimation-based Distributed Output Regulation

Let a group of N LTI agents be described by the differential equation

$$\dot{x}_k = A_k x_k + B_k u_k + B_k^w w_k + B_k^\xi \xi + B_k^\zeta \zeta_k, \tag{4.89}$$

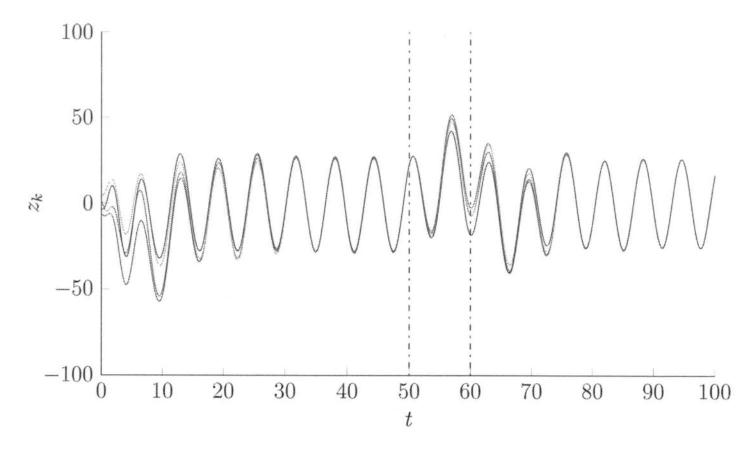

Figure 4.6: Synchronizing behaviour of the output z_k.

where for all $k \in \mathcal{N}$, $x_k(t) \in \mathbb{R}^{n_k}$ is the state of the agent, $u_k(t) \in \mathbb{R}^{m_k}$ is the respective control input, and w with $w_k(t) \in \mathbb{R}^{v_k}$ is the $\mathcal{L}_2$-integrable disturbance affecting agent k. Moreover, the agents are disturbed by the exosystem states ξ and ζ_k.

Here, $\xi \in \mathbb{R}^v$ is the state of the global exosystem

$$\dot{\xi} = S\xi, \tag{4.90}$$

and $\zeta_k \in \mathbb{R}^{\tilde{v}_k}$ the state of the local exosystems

$$\dot{\zeta}_k = \widetilde{S}_k\zeta_k. \tag{4.91}$$

Remark 4.9 *Basically,* (4.90) *and* (4.91) *can also be formulated as a single exosystem without distinguishing between a global and a local exosystem. However, splitting the exosystem as seen above is meaningful, in order to restrict the estimates the relevant parts.*

The measurement output y_k that agent k receives is defined as

$$y_k = C_k x_k + \sum_{j \in \mathcal{N}_k} C_{kj} x_j + D_k^{\xi}\xi + D_k^{\zeta}\zeta_k + \eta_k, \tag{4.92}$$

where η_k with $\eta_k(t) \in \mathbb{R}^{r_k}$ is a $\mathcal{L}_2$-integrable measurement disturbance function. Now, in addition to the definition in the previous section, we consider the case where y_k is influenced by the exosystem states ξ and ζ_k. Using this definition, many kinds of disturbances on y_k can be modelled such as constant offsets and oscillating forces. Moreover, it can be distinguished whether these disturbances affect the agents individually or the complete MAS.

The regulation outputs of agent k, $k \in \mathcal{N}$, are then defined as

$$z_k = C_k^z x_k + E_k^\xi \xi + E_k^\zeta \zeta_k \tag{4.93}$$

and the output regulation problem can be defined as follows:

Problem 7 (Distributed output regulation) *For every agent k, determine a dynamic regulator, which receives the measurement y_k and data communicated from the neighbouring agents $j \in \mathcal{N}_k$, such that:*

1. *In the absence of exogenous and measurement disturbances (i.e., when $w_k = 0$, $\eta_k = 0$, for all $k \in \mathcal{N}$)*
$$\lim_{t\to\infty} z_k(t) = 0, \quad k \in \mathcal{N}.$$

2. *$\mathcal{H}_\infty$-type regulation performance is guaranteed in the sense that*
$$\sum_{k=1}^{N} \int_0^\infty z_k^\top z_k dt \leq \kappa^2 \sum_{k=1}^{N} \int_0^\infty \|w_k\|^2 dt + \theta^2 \sum_{k=1}^{N} \int_0^\infty \|\eta_k\|^2 dt + I_0 \tag{4.94}$$
with performance parameter κ and θ and the initial costs $I_0 \geq 0$ due to the transient regulation error, which depends on the initialization of the agents and the dynamic controllers.

The outputs (4.92) are defined very generally and includes regular output stabilization tasks under disturbances, individual reference tracking, as well as leader-follower synchronization, where in the latter case, the global exosystem would act as the leader. Moreover, additional coordination schemes can be implemented using (4.93), where for instance, certain relative positions of the agents with respect to the global reference are desired.

In order to design the regulators, as the first step, we implement the regional estimation scheme from Section 4.2.1, where some extension is needed due to the existence of the global and local exosystems. To be precise, the individual estimators are assigned to also estimate the global exosystem's state ξ and the respective local exosystems' states $\zeta_j, j \in \mathcal{N}_k$.

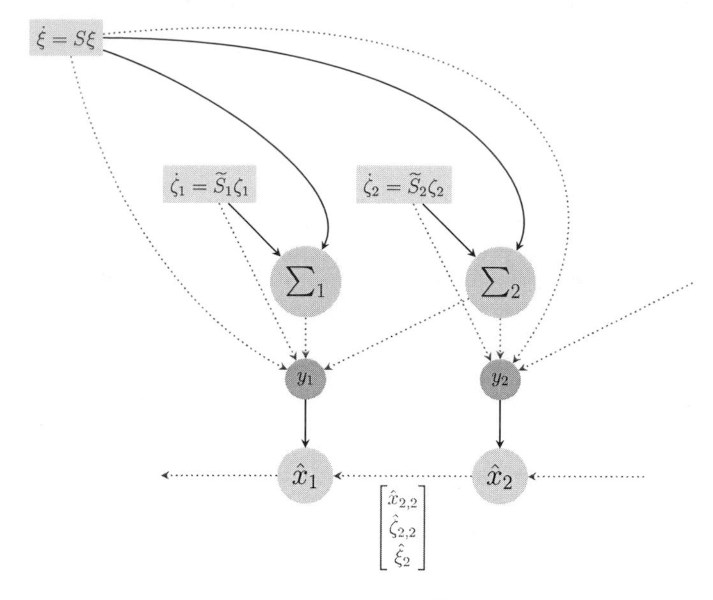

Figure 4.7: Illustration of the interactions, where $\sum_k$ represents the agent k. Dotted arrows show communication and algebraic dependencies, solid arrows show feedback or feedforward interaction.

The estimates $\hat{x}^k$ and the estimation error vectors e_k are thus defined as

$$e_k = \overline{x}_k - \hat{x}_k = \begin{bmatrix} x_k \\ \zeta_k \\ \begin{bmatrix} x_j \\ \zeta_j \end{bmatrix}_{j\in\mathcal{N}_k} \\ \xi \end{bmatrix} - \begin{bmatrix} \hat{x}_{k,k} \\ \hat{\zeta}_{k,k} \\ \begin{bmatrix} \hat{x}_{k,j} \\ \hat{\zeta}_{k,j} \end{bmatrix}_{j\in\mathcal{N}_k} \\ \hat{\xi}_k \end{bmatrix}.$$

Figure 4.7 illustrates the interactions between the subsystems. The observer differential equations are proposed as

$$\dot{\hat{x}}_k = \overline{A}_k\hat{x}_k + L_k(y_k - \overline{C}_k\hat{x}_k) + K_k \begin{bmatrix} \begin{bmatrix} \hat{x}_{j,j} - \hat{x}_{k,j} \\ \hat{\zeta}_{j,j} - \hat{\zeta}_{k,j} \end{bmatrix}_{j\in\mathcal{N}_k} \\ \frac{1}{p_k}\sum_{j\in\mathcal{N}_k}(\hat{\xi}_j - \hat{\xi}_k) \end{bmatrix} + \overline{B}_k\overline{u}_k, \tag{4.95}$$

where we define the matrices

$$\overline{A}_k = \begin{bmatrix} A_k & B_k^\zeta & 0 & B_k^\xi \\ 0 & \widetilde{S}_k & 0 & 0 \\ 0 & 0 & \operatorname{diag}\begin{bmatrix} A_j & B_j^\zeta \\ 0 & \widetilde{S}_j \end{bmatrix}_{j\in\mathcal{N}_k} & \begin{bmatrix} B_j^\xi \\ 0 \end{bmatrix}_{j\in\mathcal{N}_k} \\ 0 & 0 & 0 & S \end{bmatrix}, \quad \overline{C}_k = \begin{bmatrix} C_{kk}^\top \\ D_k^{\zeta\top} \\ \begin{bmatrix} C_{kj}^\top \\ 0 \end{bmatrix}_{j\in\mathcal{N}_k} \\ D_k^{\xi\top} \end{bmatrix}^\top.$$

The observers may be initialized arbitrarily as $\hat{x}_k(0) = \hat{x}_0^k \in \mathbb{R}^{\sigma_k}$. Moreover, we define the input matrices

$$\overline{B}_k = \begin{bmatrix} B_k & 0 \\ 0_{\tilde{v}_k\times m_k} & 0 \\ 0 & \operatorname{diag}\begin{bmatrix} B_j \\ 0_{\tilde{v}_j\times m_j} \end{bmatrix}_{j\in\mathcal{N}_k} \\ 0 & 0 \end{bmatrix}, \quad B_k^w = \begin{bmatrix} B_k^d & 0 \\ 0_{\tilde{v}_k\times \tilde{m}_k} & 0 \\ 0 & \operatorname{diag}\begin{bmatrix} B_j^d \\ 0_{\tilde{v}_j\times \tilde{m}_j} \end{bmatrix}_{j\in\mathcal{N}_k} \\ 0 & 0 \end{bmatrix}.$$

As in Section 4.2.1, the observer gain matrices L_k and K_k need to be determined. Moreover, the observers (4.95) uses the combined inputs $\overline{u}_k$, c.f. Remark 4.4, which is undesirable and will be addressed later.

Remark 4.10 *The system class considered here* (4.89), (4.92) *generalized the one considered in (Seyboth et al., 2016). In fact, by assuming $C_{kj} = 0$ for $k \neq j$ and detectability of*

$$\left(\begin{bmatrix} A_1 & B_1^\zeta & B_1^\xi \\ 0 & \widetilde{S}_1 & 0 \\ 0 & 0 & S \end{bmatrix}, \begin{bmatrix} C_{11} & D_1^\zeta & D_1^\xi \end{bmatrix} \right) \text{ and } \left(\begin{bmatrix} A_k & B_k^\zeta \\ 0 & \widetilde{S}_k \end{bmatrix}, \begin{bmatrix} C_k & D_k^\zeta \end{bmatrix} \right), k \in \mathcal{N}\setminus 1$$

in (4.89), (4.92), *we recover the class considered in (Seyboth et al., 2016). With these assumptions, it is sufficient for the agents to merely transmit the estimates of the global exosystems $\hat{\xi}^k$. The more general system class considered here introduces a higher demand for communication by definition of the distributed observers* (4.95).

Designing the observer gain matrices can now be done in similar fashion to Section 4.2.1. We define the matrices

$$Q_k = P_k\overline{A}_k + \overline{A}_k^\top P_k - 2G_k\overline{C}_k - F_k\begin{bmatrix} 0_{\tau_k\times(n_k+\tilde{v}_k)} I_{\tau_k}\end{bmatrix} - (F_k\begin{bmatrix} 0_{\tau_k\times(n_k+\tilde{v}_k)} I_{\tau_k}\end{bmatrix})^\top + \alpha_k P_k + q_k\pi_k \begin{bmatrix} P_{k,11} & 0 & 0 \\ 0 & 0_{(\tau_k-v)\times(\tau_k-v)} & 0 \\ 0 & 0 & P_{k,v} \end{bmatrix},$$

where $P_k \succ 0, G_k$, and F_k are solution variables and $\beta > 0$ is to be minimized. Moreover, $P_{k,11}$ defines the principal submatrix of P_k with the first $n_k + \widetilde{v}_k$ row/columns, and $P_{k,v}$ defines the principal submatrix of P_k with the last v rows/columns; $\pi_k, \alpha_k > 0$ are scalar parameters in analogy to section 4.2.1.

Theorem 4.9 *Consider a group of N agents defined by* (4.89),(4.92) *for all $k \in \mathcal{N}$ and suppose Assumption 4.1 holds. For given weighting matrix W_k, let a collection of matrices F_k, $P_k \succ 0$, $k \in \mathcal{N}$, and a performance parameter $\beta \geq 0$ be a solution of the LMIs*

$$\left[\begin{array}{ccc|c} Q_k + W_k & -G_k & P_k B^{d,k} & F_k \begin{bmatrix} I_{\tau_k - v} & 0 \\ 0 & \frac{1}{p_k}\mathbf{1}_{p_k}^\top \otimes I_v \end{bmatrix} \\ -G_k t & -\beta I & 0 & 0 \\ (P_k B^{d,k})^\top & 0 & -\beta I & 0 \\ \hline \begin{bmatrix} I_{\tau_k - v} & 0 \\ 0 & \frac{1}{p_k}\mathbf{1}_{p_k} \otimes I_v \end{bmatrix} F_k t & 0 & 0 & -\begin{bmatrix} [\pi_j P_{11}^j]_{j \in \mathcal{N}_k} & 0 \\ 0 & [\pi_j P_v^j]_{j \in \mathcal{N}_k} \end{bmatrix} \end{array}\right] \preceq 0 \tag{4.96}$$

for all $k \in \mathcal{N}$. Then, the distributed observers (4.95) *with the gain matrices*

$$\begin{aligned} L_k &= (P_k)^{-1} G_k, \\ K_k &= (P_k)^{-1} F_k, \end{aligned} \tag{4.97}$$

satisfy

1. *In the nominal case where $w_k \equiv 0, \eta_k \equiv 0$ for all $k \in \mathcal{N}$,*

$$\lim_{t \to \infty} e^k(t) = 0, \quad k \in \mathcal{N}. \tag{4.98}$$

2. *The estimation errors e_k satisfy $\mathcal{H}_\infty$-type performance in the sense that*

$$\sum_{k=1}^{N} \int_0^\infty e^{k\top} W_k e^k dt \leq \beta \sum_{k=1}^{N} \int_0^\infty (\|w_k\|^2 + \|\eta_k\|^2) dt + J_0 \tag{4.99}$$

Proof. The proof follows the lines of the proof of Theorem 4.4. ■

The discussions on the distributed observers in Section 4.2.1 concerning convergence speed and solution methods are also applicable here. Also, the observers (4.95) use the input vector $\overline{u}_k$, which includes the neighbour inputs as already seen in Section 4.2.1. Subsequently, the same approach as shown in Corollary 4.1 can be used to eliminate this issue.

Now, as a suitable estimation scheme is available, we next focus on the construction of state feedback regulators that are able to solve Problem 7. Note that in order to solve Problem 7, designing the control law and the observers are not independent, as already encountered in Sections 4.2, 4.3.1, and 4.3.2.

Since in the present problem, we consider separate global and local exosystems, the Regulator Equations (4.5) can also be decomposed, which is shown in (Seyboth et al., 2016). The Regulator Equations therefore take the form

$$\begin{aligned} &A_k\Pi_k + B_k\Gamma_k - \Pi_k S + B_k^{\xi} = 0,\\ &C_k^{e}\Pi_k + E_k^{\xi} = 0,\\ \\ &A_k\widetilde{\Pi}_k + B_k\widetilde{\Gamma}_k - \widetilde{\Pi}_k\widetilde{S}_k + B_k^{\zeta} = 0,\\ &C_k^{e}\widetilde{\Pi}_k + E_k^{\zeta} = 0, \end{aligned} \tag{4.100}$$

which we assume to have a solution for every $k \in \mathcal{N}$. Then, let the control law be defined as

$$u_k = \Gamma_k\hat{\xi}_k + \widetilde{\Gamma}_k\hat{\zeta}_k^k + H_k(\Pi_k\hat{\xi}_k + \widetilde{\Pi}_k\hat{\zeta}_k^k - \hat{x}_{k,k}), \tag{4.101}$$

which gives us following result on the design of the closed loop controllers.

Theorem 4.10 *Let a group of N agents be given as* (4.89) *with regulation outputs* (4.93) *and measurement outputs* (4.92)*, and let the communication topology be defined by a graph $\mathcal{G} = (\mathcal{V}, \mathcal{E})$. Let Assumption 4.1 and 4.2 be satisfied and suppose the following conditions hold for all $k \in \mathcal{N}$:*

1. *There exist matrices $\Pi_k, \widetilde{\Pi}_k, \Gamma_k$, and $\widetilde{\Gamma}_k$, which solve (4.100).*

2. *There exists $X_k > 0$, such that $A_k - \frac{1}{\lambda_k^2}B_kB_k^\top X_k$ is Hurwitz, and the Algebraic Riccati Equation*

$$X_kA_k + A_k^\top X_k + C_k^{e\top}C_k^{z} + \widetilde{\alpha}_k I_{n_k} - X_k\left(\frac{1}{\lambda_k^2}B_kB_k^\top - \frac{1}{\mu_k^2}B_k^{w}B_k^{d\top}\right)X_k = 0 \tag{4.102}$$

   is solved with $\mu_k, \lambda_k > 0$, $\widetilde{\alpha}_k \geq 0$. The parameter $\widetilde{\alpha}_k$ may be arbitrarily small, but $\widetilde{\alpha}_k > 0$ if (A_k, C_k^z) is undetectable.

3. *The LMI* (4.96) *is solved with the weighting matrix*

$$W_k = \begin{bmatrix} \frac{3}{\lambda_k^2}X_kB_kB_k^\top X_k & 0 & 0 & 0\\ 0 & 3\lambda_k^2\widetilde{T}_k^\top\widetilde{T}_k & 0 & 0\\ 0 & 0 & 0_{(\tau_k-v)\times(\tau_k-v)} & 0\\ 0 & 0 & 0 & 3\lambda_k^2T_k^\top T_k \end{bmatrix}, \tag{4.103}$$

where

$$\begin{aligned}\widetilde{T}_k &= \widetilde{\Gamma}_k + \frac{1}{\lambda_k^2} B_k^\top X_k \widetilde{\Pi}_k \\ T_k &= \Gamma_k + \frac{1}{\lambda_k^2} B_k^\top X_k \Pi_k\end{aligned}. \tag{4.104}$$

Then, Problem 7 is solved with the distributed observers (4.95) *with the observer gain matrices* (4.97) *and the control law* (4.101) *with*

$$H_k = \frac{1}{\lambda_k^2} B_k^\top X_k. \tag{4.105}$$

The guaranteed performance is given by $\kappa = \sqrt{\max_k \mu_k^2 + \beta}$ *and* $\theta = \sqrt{\beta}$.

Proof. The proof uses similar techniques as the proof of Theorem 4.5. Its complete version is given in Appendix B.4. ■

The items in Theorem 4.10 present the design procedure of the distributed regulators. In particular, the Algebraic Ricatti Equation may be solved in purely decentralized fashion. It should be noted that in this section, we did not assume the communication graph to be connected. The reason for that is the character of the problem, where the local exosystem may present individual references for the agents, so no connectedness of $\mathcal{G}$ may be needed. Concerning the choice of the parameters λ_k and μ_k, Remark 4.6 is applicable here.

Simulation Example:

In order to demonstrate the presented output regulators, we simulate five omnidirectional vehicles which are modelled as double integrators with two degrees of freedom. Their position shall be regulated to a fixed distance with respect to a global reference trajectory y_{ref}, which is given by an oscillating global exosystem (4.106). The reference outputs y_{ref} are illustrated in Figure 4.8.

$$\begin{aligned}\dot{\xi} &= \begin{bmatrix} 0 & 1 & 0 & 1 & 0 & 1 \\ -1 & 0 & 1 & 0 & 1 & 0 \\ 0 & -1 & 0 & 1 & 0 & 1 \\ -1 & 0 & -1 & 0 & 1 & 0 \\ 0 & -1 & 0 & -1 & 0 & 1 \\ -1 & 0 & -1 & 0 & -1 & 0 \end{bmatrix} \xi \\ y_{ref} &= \begin{bmatrix} -1 & 0 & 0 & 0 & 0 & 0 \\ 0 & -1 & 0 & 0 & 0 & 0 \end{bmatrix} \xi.\end{aligned} \tag{4.106}$$

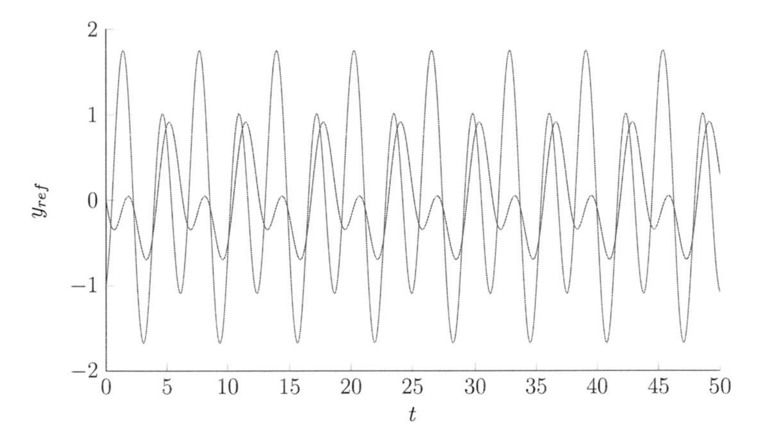

Figure 4.8: Reference trajectory defined by the global exosystem.

The assigned distances to y_{ref} are assumed to be constant and thus are implemented by local exosystems with constant states. Moreover, the agents $k = 1, ..., 3$ are affected by oscillatory input disturbances, which are also generated by the local exosystems (4.107).

$$k = 1,2,3: \quad \dot{\zeta}_k = \begin{bmatrix} 0 & 0 & 0 & 0 \\ 0 & 0 & 0 & 0 \\ 0 & 0 & 0 & 1 \\ 0 & 0 & -1 & 0 \end{bmatrix} \zeta_k \qquad k = 4,5: \quad \dot{\zeta}_k = \begin{bmatrix} 0 & 0 \\ 0 & 0 \end{bmatrix} \zeta_k \tag{4.107}$$

The agent differential equations are given as

$$\dot{x}_k = \begin{bmatrix} 0 & 1 & 0 & 0 \\ 0 & 0 & 0 & 0 \\ 0 & 0 & 0 & 1 \\ 0 & 0 & 0 & 0 \end{bmatrix} x_k + \begin{bmatrix} 0 & 0 \\ 1 & 0 \\ 0 & 0 \\ 0 & 1 \end{bmatrix} u_k + \begin{bmatrix} 0 & 0 \\ 0.1 & 0 \\ 0 & 0 \\ 0 & 0.1 \end{bmatrix} w_k + B_k^\zeta \zeta_k,$$

where

$$k = 1,2,3: \quad B_k^\zeta = \begin{bmatrix} 0 & 0 & 0 & 0 \\ 0 & 0 & 1 & 0 \\ 0 & 0 & 0 & 0 \\ 0 & 0 & 0 & 1 \end{bmatrix} \qquad k = 4,5: \quad B_k^\zeta = \begin{bmatrix} 0 & 0 \\ 0 & 0 \\ 0 & 0 \\ 0 & 0 \end{bmatrix}$$

and w_k are $\mathcal{L}_2$-integrable exogeneous disturbances like wind or slopes. The regulation outputs are defined as

$$z_k = \underbrace{\begin{bmatrix} 1 & 0 & 0 & 0 \\ 0 & 0 & 1 & 0 \end{bmatrix}}_{C_k^e} x_k + \underbrace{\begin{bmatrix} 1 & 0 & 0 & 0 & 0 & 0 \\ 0 & 1 & 0 & 0 & 0 & 0 \end{bmatrix}}_{z_k^\xi} \xi + E_k^\zeta \zeta_k, \tag{4.108}$$

with

$$k = 1,2,3: \quad E_k^\zeta = \begin{bmatrix} 1 & 0 & 0 & 0 \\ 0 & 1 & 0 & 0 \end{bmatrix} \qquad k = 4,5: \quad E_k^\zeta = \begin{bmatrix} 1 & 0 \\ 0 & 1 \end{bmatrix}.$$

Let the communication topology be given by the directed ring-graph with

$$\mathcal{E} = \{(v_2, v_1), (v_3, v_2), (v_4, v_3), (v_5, v_4), (v_1, v_5)\}, \tag{4.109}$$

where the agents measure the relative position w.r.t. their neighbour. The measurements of agent 4 and 5 are moreover disturbed by the global exosystem. In addition, agent 1 and 3 obtain some relative measurement w.r.t. the global exosystem. Thus, the measurement outputs are given as

$$y_1 = \begin{bmatrix} Cx_2 - Cx_1 \\ \begin{bmatrix} 1 & 0 & 0 & 0 \\ 0 & 1 & 0 & 0 \end{bmatrix} \zeta_1 \\ \begin{bmatrix} 1 & 0 \end{bmatrix} (D^\xi \xi - Cx_1) \end{bmatrix} \quad y_2 = \begin{bmatrix} Cx_3 - Cx_2 \\ \begin{bmatrix} 1 & 0 & 0 & 0 \\ 0 & 1 & 0 & 0 \end{bmatrix} \zeta_2 \end{bmatrix} \quad y_3 = \begin{bmatrix} Cx_4 - Cx_3 \\ \begin{bmatrix} 1 & 0 & 0 & 0 \\ 0 & 1 & 0 & 0 \end{bmatrix} \zeta_3 \\ \begin{bmatrix} 0 & 1 \end{bmatrix} (D^\xi \xi - Cx_1) \end{bmatrix}$$

$$y_4 = \begin{bmatrix} Cx_5 - Cx_4 + D^\xi \xi \\ \zeta_4 \end{bmatrix} \quad y_5 = \begin{bmatrix} Cx_1 - Cx_5 + D^\xi \xi \\ \zeta_5 \end{bmatrix} \quad D^\xi = \begin{bmatrix} 0 & 0 & 0 & 0 & 1 & 0 \\ 0 & 0 & 0 & 0 & 0 & 1 \end{bmatrix}$$

with $C = \begin{bmatrix} 1 & 0 & 0 & 0 \\ 0 & 0 & 1 & 0 \end{bmatrix}$. The Regulator equations (4.100) are solved by

$$k \in \mathcal{N}: \quad \Pi_k = \begin{bmatrix} 1 & 0 & 0 & 0 & 0 & 0 \\ 0 & 1 & 0 & 1 & 0 & 1 \\ 0 & 1 & 0 & 0 & 0 & 0 \\ -1 & 0 & 1 & 0 & 1 & 0 \end{bmatrix}, \qquad \Gamma_k = \begin{bmatrix} -3 & 0 & -1 & 0 & 1 \\ 0 & -3 & 0 & -1 & 0 \end{bmatrix}$$

$$k = 1,2,3: \quad \widetilde{\Pi}_k = \begin{bmatrix} -1 & 0 & 0 & 0 \\ 0 & 0 & 0 & 0 \\ 0 & -1 & 0 & 0 \\ 0 & 0 & 0 & 0 \end{bmatrix}, \qquad \widetilde{\Gamma}_k = \begin{bmatrix} 0 & 0 & -1 & 0 \\ 0 & 0 & 0 & -1 \end{bmatrix}$$

$$k = 4,5: \quad \widetilde{\Pi}_k = \begin{bmatrix} -1 & 0 \\ 0 & 0 \\ 0 & -1 \\ 0 & 0 \end{bmatrix}, \qquad \widetilde{\Gamma}_k = \begin{bmatrix} 0 & 0 \\ 0 & 0 \end{bmatrix}.$$

With the Algebraic Riccati Equations (4.102) and the parameters $\lambda_k, \mu_k = 0.1, \widetilde{\alpha}_k = 0$, we obtain feedback gain matrices of

$$H_k = \begin{bmatrix} 10.05 & 4.51 & 0 & 0 \\ 0 & 0 & 10.05 & 4.51 \end{bmatrix} \qquad k \in \mathcal{N}.$$

The observer gain matrices L_k and K_k are obtained by solving the LMIs (4.96), where we have $\beta = 3.46$ by minimization. This solves Problem 7 with $\kappa = 1.863$ and $\theta = 1.860$. Figure 4.9 shows the positions of the agents.

4.4 Summary and Discussion

In this chapter, we presented solutions to the synchronization and output regulation problems for multi-agent systems with coupled measurements, which are foremost motivated from groups of vehicle that have sensors measuring relative positions instead of their absolute coordinates. However, potential application reach well beyond that to e.g. flexible structures, where only relative positions are measured, or heterogeneous population models, where sensors cannot distinguish between the subpopulations, but can only measure the size of the total population.

The two approaches that we presented are fundamentally different, as in the first approach, we aim at homogenization of the agents by expansion of the persistent dynamics, and in the second approach, we aim at making use of heterogeneity of the agents to design distributed observers that can deliver the agents' absolute states.

The first approach, expansion of the models, requires some geometric conditions and does not allow for a lot of freedom in assigning an desired synchronized trajectory, since the Internal Model dynamics, respectively S, is determined by the agents' eigenvalues on the imaginary axis.

The second approach, applying distributed estimation also requires some detectability properties as discussed in this Chapter. However, once the estimation problem is solved, it allows for more flexibility in choosing S, and thus actively shaping the synchronized trajectory.

In order to preserve scalability of the resulting observer, the distributed estimation scheme from Chapter 2 needs to be modified. We introduced this modification as *re-*

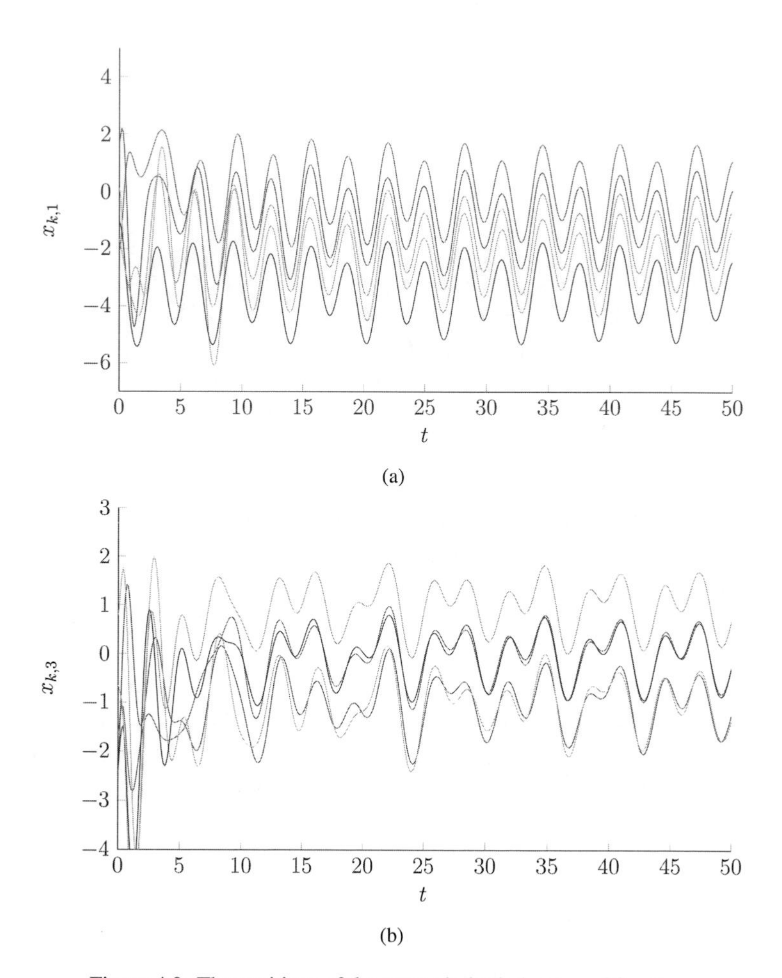

Figure 4.9: The positions of the agents in both degrees of freedom.

gional estimation, where every observer estimates the local agent's state and those of its neighbours. The name *regional estimation* was chosen as a middle-ground between global and local estimation. This way, scalability of the resulting observers is preserved, as the observer dimension do not grow with the size of the complete network. The observer gain matrices are again found by solving distributed LMI-conditions, while providing for $\mathcal{H}_\infty$-type performance guarantees of the estimates with respect to exogeneous disturbances and measurement disturbances. A similar approach was pursued in (Wahrburg & Adamy, 2011; Listmann et al., 2011), where however the observer do not cooperate and thus require much stricter detectability conditions. Our estimation scheme on the other hand combines the benefits of both local and global estimation: scalability, performance, and efficient design of the observers.

Even though we introduced this estimation scheme as a way to solve the synchronization task, it is of general value: consider the distributed estimation of unstructured systems from Chapter 2. If every observer is assigned to estimate solely a partial state x_k of x of lower dimension, then our methods from *regional estimation* can be applied. Physical interconnection of x_k with $x_j, j \neq k$, however, need to be properly addressed in this case, which is discussed as one of the extensions in this chapter. Two more extensions are presented, first relaxing the requirement on the knowledge about the neighbours' models, and secondly extending the system class by including exosystems that generate disturbances or reference signals.

Note that in our estimation-based synchronization/output regulation results, we are able to guarantee some closed-loop $\mathcal{H}_\infty$-type performance of the error function with respect to exogenous disturbances and measurement disturbances. To the best of our knowledge, this is something that has not been done before for heterogeneous MAS, and especially not for the general case of coupled measurements. Moreover, we showed that our regional estimation scheme can contribute positively towards the performance guarantees.

Chapter 5

Conclusions

The starting point of this thesis is the distributed estimation scheme (2.6), where the observer gain matrices are designed by distributed LMI-conditions (2.21). From there on, we developed extensions in a wide range of directions.

Before wrapping up our contributions, we revisit the literature on networked control in general. Since this field started to grow rapidly over ten years ago, three main *directions of complexity* have crystallized:

- **System complexity:** Going from single-integrators, via linear systems, to linear parameter/time-varying systems or general nonlinear systems, etc.
- **Communication complexity:** Going from perfect communication channels, to realistic communication channels with communication delays, package dropout, bandwith-limitations, etc.
- **Topological complexity:** Going from unstructured/single systems, via fixed topologies, to randomly varying topologies, etc.

The directions of communication complexity and system complexity are particularly important to enhance applicability of networked control results, while topological complexity needs to be addressed when the system to be controlled is a multi-agent systems. Reviewing our literature overview from Section 1.3 and 2.3, one notices that essentially all publications follow one of these directions of complexity, or combine multiple directions. Thus, it is fair to categorize these publications in the form of a *complexity cube*, which is a notion that the *Institute for Systems Theory and Automatic Control* at the *University of Stuttgart* has been using for several years, see e.g. (Wieland, 2010).

In analogy to the complexity cube of networked control in general, for this thesis, we span the complexity cube for distributed estimation specifically, with the same denomina-

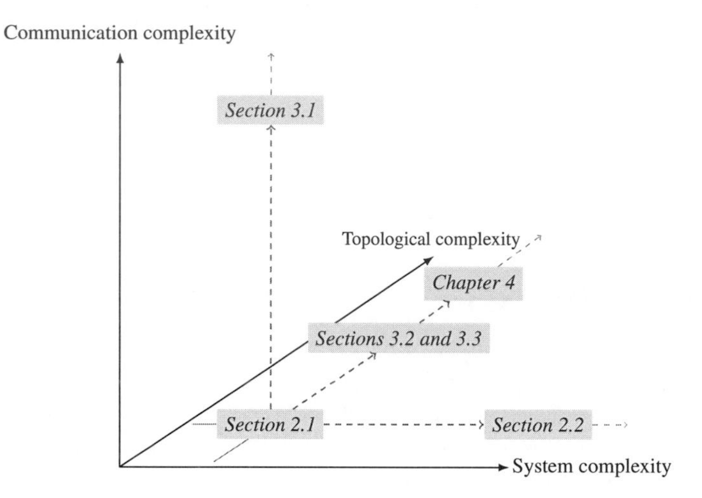

Figure 5.1: Contributions of this thesis in the complexity cube

tion of the three axis as above. Figure 5.1 shows the categorization of our results and as it can be seen, we have expanded the theory of distributed estimation in all three directions.

- In Section 2.2, our results focus on system complexity by introducing additive nonlinearities into the system description and adapting the distributed observers such that they can handle a large class, namely incrementally quadratically constrained nonlinearities. Thereby, we proposed two approaches for observer design, where the first one is based on a vector-Lyapunov-function approach, which only requires rather simple LMI-conditions, but has proven to be restrictive in the case of non-globally-Lipschitz nonlinearities. The second approach requires a two-step optimization approach and includes more complex LMI-conditions, but can handle systems that the first approach cannot.

  Studying incrementally quadratically constrained nonlinearities surely is not the end of the line when it comes to extending system complexity. As already discussed in the summary of Chapter 2, considering methods using multipliers (Zames & Falb, 1968; Materassi & Salapaka, 2011) or considering nonlinearity satisfying integral quadratic constraints (Jonsson & Megretski, 2000; Veenman & Scherer, 2011), may yield considerable extension.

- Communication complexity was addressed in Section 3.1, where we introduced an event-triggered scheme to avoid continuous communication and thus reducing transmissions. We proved that our triggering scheme is well-behaved in the sense that Zeno-behaviour cannot happen. Furthermore, we established a direct relation between the guaranteed minimum inter-event time and parameters, where some are specific to the system and some are design parameters. This opens opportunities for tuning the distributed estimators such that observer states need to be transmitted as rarely as possible.

  As already discussed in the summary of Chapter 2 and 3, reducing the required communication bandwidth is not the only issue. In digital communication, communication delays and occasional package losses are likely to happen, which has been thoroughly discussed e.g. in (Schenato et al., 2007; Schenato, 2008), to name a few, and therefore, this may also be considered in distributed estimation schemes. However, in the case of communication delays, we expect that it may severely add complexity to the LMI-conditions for observer design. In particular, the design freedom in defining a Lyapunov-Krasovskii functional requires balancing conservativeness versus complexity of the LMI-condition, which can be often observed in $\mathcal{H}_\infty-$type control/estimation of time-delay systems, e.g. (Fridman & Shaked, 2001). Furthermore, investigating in periodic communication and self-triggered communication may add more ways to address bandwidth-issue (X. Wang & Lemmon, 2009; Anta & Tabuada, 2010), so that in application, a large variety of communication protocols can be chosen from.

- Topological complexity in distributed estimation is considered as two different types: First, we have observer communication topology. Here, one important problem that we addressed was the distributed design of the observer gain matrices, presented in Sections 3.2, 3.3. Secondly, for large-scale systems, agent topology becomes relevant, as for growing size of an MAS; it is desirable for the observers to keep a small dimension. To address this issue, we presented a regional estimation scheme in Chapter 4 and applied it to the synchronization and output regulation problem of heterogeneous MAS.

As this recap shows, we have presented a number of novel extension to the solution of the distributed estimation problem and thus enhanced applicability with respect to the system class, agent topology, computation of observer gain matrices, and communication channels. In every direction of complexity, the results can be further improved or more generality can be achieved, as we discussed here. Furthermore, our results in this thesis present the extensions separately in the three directions of complexity. An unifying framework may be highly interesting and subject to future work.

Appendix A

Technical Preliminaries

In this section, we aim at giving some basics on observer design and graph theory, which are assumed to be known in the main part of this thesis. This is only a short review and focusses on concepts without presenting much detail.

A.1 Observer design

A.1.1 Undisturbed case

Let us first consider the undisturbed autonomous system in state space representation

$$\dot{x} = f(x) + g(x, u), \tag{A.1}$$

where x is the n-dimensional state vector and u is the m-dimensional control input vector, i.e. $x(t) \in \mathbb{R}^n, u(t) \in \mathbb{R}^m$. It is assumed that certain measurements of the system (A.1) can be obtained in order to estimate its state x. Those measurements are assumed to be given in the form of an r-dimensional output vector

$$y = h(x, u). \tag{A.2}$$

Moreover, it is assumed that also the control input u is known. The observer design problem is now defined as creating a dynamical system

$$\dot{\hat{x}} = \hat{f}(\hat{x}, y, u), \tag{A.3}$$

such that

$$\lim_{t \to \infty} (\hat{x}(t) - x(t)) = 0. \tag{A.4}$$

In particular, in the case of linear systems

$$\begin{aligned}\dot{x} &= Ax + Bu,\\ y &= Cx + Du,\end{aligned} \tag{A.5}$$

the solution to the observer design problem is well studied in literature and an efficient solution method was introduced in (Luenberger, 1966), where the observer is designed as

$$\hat{x} = A\hat{x} + L(y - C\hat{x} - Du) + Bu. \tag{A.6}$$

Here, $L \in \mathbb{R}^{n\times r}$ is the observer gain matrix that is to be designed, which can be done by placing the poles of the matrix $A - LC$ in the open left complex place $\mathbb{C}^-$, thus rendering $A - LC$ Hurwitz.

The case of nonlinear systems (A.1) also has seen many different approaches for observer design. Among the first to be introduced was the robustness-based approach (Thau, 1973). There, nonlinear systems are considered which can be decomposed into a linear systems and an additive nonlinearity

$$\begin{aligned}\dot{x} &= Ax + f(x) + Bu,\\ y &= Cx + Du,\end{aligned} \tag{A.7}$$

where it is additionally assumed that $f(x)$ satisfies a global Lipschitz condition, i.e.

$$\|f(a) - f(b)\| \leq \theta\|a - b\| \quad \text{for all } a, b \in \Omega, \tag{A.8}$$

with $\theta > 0$ and some region $\Omega \subset \mathbb{R}^n$ such that $x(t) \in \Omega$ for all t. The observer is given as

$$\hat{x} = A\hat{x} + f(\hat{x}) + L(y - C\hat{x} - Du) + Bu. \tag{A.9}$$

Subsequently, the nonlinearity is treated as a disturbance, whose diminishing effect on the estimate is suppressed by a large observer gain matrix.

In the following, researchers have developed various extensions. For instance, efficient methods for calculating the observer gain matrices where presented in (Raghavan & Hedrick, 1994) and methods for extensions on the generalization of the global Lipschitz condition (A.8) where presented in (Arcak & Kokotović, 2001b), (Açikmese & Corless, 2011).

A.1.2 $\mathcal{L}_2$-integrable disturbances

In the case of $\mathcal{L}_2$-integrable disturbances, $\mathcal{H}_\infty$-performance can be guaranteed in the sense of

$$\int_{\tau=0}^{t} e^\top W e d\tau < \gamma^2 \int_{\tau=0}^{t} (w^\top w + \eta^\top \eta) d\tau \tag{A.10}$$

for the $\mathcal{H}_\infty$-gain of $\gamma > 0$ to be minimized. The observer gain matrices are calculated by solving the LMI

$$\left[\begin{array}{c|cc} PA + A^\top P + GC + C^\top G^\top + W & PB & G \\ \hline B^\top P & -\gamma^2 I & 0 \\ G^\top & 0 & -\gamma^2 I \end{array}\right] \preceq 0$$

and setting $L = P^{-1}G$. This result can be easily obtained by considering the storage function candidate $V = e^\top Pe$ and is closely related to results from $\mathcal{H}_\infty$-control (Skogestad & Postlethwaite, 2007).

A.2 Graph Theoretic Preliminaries

In this section we briefly introduce some definitions on communication graphs. For a thorough discourse, the reader is referred to (Mesbahi & Egerstedt, 2010). We use directed, unweighted graphs $\mathcal{G} = (\mathcal{V}, \mathcal{E})$ to describe the interconnections between the individual observers: $\mathcal{V} = \{v_1, ..., v_N\}$ is the set of vertices, where $v_k \in \mathcal{V}$ represents the k-th observer; $\mathcal{E} \subset V \times V$ is the set of edges, which models information flow, i.e. the k-th agent receives information from the j-th agent if and only if $(v_j, v_k) \in \mathcal{E}$. The neighbourhood of observer k is denoted $\mathcal{N}_k = \{j | (v_j, v_k) \in \mathcal{E}\}$ and the outgoing-neighbourhood is denoted $\mathcal{M}_k = \{j | (v_k, v_j) \in \mathcal{E}\}$. A *path* from vertex v_{i_1} to vertex v_{i_l} is a sequence of vertices $\{v_{i_1}, ..., v_{i_l}\}$ such that $(v_{i_j}, v_{i_{(j+1)}}) \in \mathcal{E}, j = 1, ..., l-1$. Moreover, we define the in-degree and out degree of observer k as $p_k = |\mathcal{N}_k|$ and $q_k = |\mathcal{M}_k|$, respectively.

Definition A.1 *A graph $\mathcal{G}$ is* undirected *or* bidirectional, *if $(v_j, v_k) \in \mathcal{E}$ implies that $(v_k, v_j) \in \mathcal{E}$.*

Definition A.2 *A graph $\mathcal{G}$ is* connected, *if there exists a vertex i, such that there exists a path from v_i to v_k, for all $k \in \mathcal{N} \backslash i$.*

Definition A.3 *A graph $\mathcal{G}$ is* strongly connected, *if for every pair of vertices v_k, v_j, $k, j = 1, ..., N, j \neq k$, there exists a path from v_k to v_j.*

In particular, the term *strongly connected* is only used if a graph is not undirected, since if a graph $\mathcal{G}$ is undirected, then the definitions of connectedness and strongly connectedness are equivalent. Also we recall the definition of independent strongly connected components defined in (Wieland, 2010).

Definition A.4 *An* independent strongly connected component (iSCC) *of a directed graph* $\mathcal{G}$ *is a subgraph* $\widetilde{\mathcal{G}} = (\widetilde{\mathcal{V}}, \widetilde{\mathcal{E}})$, *which is strongly connected and satisfies* $(v, \widetilde{v}) \notin \mathcal{E}$ *for any* $v \in \mathcal{V} \backslash \widetilde{V}$ *and* $\widetilde{v} \in \widetilde{V}$.

It holds true that every graph $\mathcal{G}$ has an iSCC and if $\mathcal{G}$ is connected, then it has exactly one iSCC.

The communication topology can also be encoded using $N \times N$-dimensional matrices of ones and zeros, introduced in the following

- $\mathcal{A}$ is the adjacency matrix, which encodes the edges with $(\mathcal{A})_{kj} = 1$ if $(v_j, v_k) \in \mathcal{E}$ and $(\mathcal{A})_{kj} = 0$ otherwise.
- $\mathcal{L}$ is the Laplacian matrix, which is $(\mathcal{L})_{kj} = -(\mathcal{A})_{kj}$ for $k \neq j$ and $(\mathcal{L})_{kk} = \sum_{j=1}^{N} (\mathcal{A})_{kj}$ for all $k \in \mathcal{N}$.

In particular, it holds that all eigenvalues of $\mathcal{L}$ have non-negative real part and $\mathcal{L}$ always has an eigenvalue 0 with the corresponding all-ones eigenvector $\mathbf{1} = [1\ 1 ... 1]^\top$. Furthermore, if $\mathcal{G}$ is connected, then $\mathbf{1}$ is the only eigenvector of the eigenvalue 0.

Appendix B

Technical proofs

B.1 Proof of Theorem 4.5

Proof. Suppose the conditions of Theorem 4.5 hold.

Now, we consider the regulation error $\epsilon_k = x_k - \Pi_k \zeta_k$. Using equations (4.49) and (4.57), the derivative of ϵ_k is found to be

$$\begin{aligned}\dot{\epsilon}_k =& \dot{x}_k - \Pi_k \dot{\zeta}_k \\ =& A_k x_k + B_k \Gamma_k \zeta_k + B_k H_k (\hat{x}_{k,k} - \Pi_k \zeta_k) + E_k w_k - \Pi_k S \zeta_k - \Pi_k \delta_k.\end{aligned}$$

It follows from (4.5) that $B_k \Gamma_k \zeta_k - \Pi_k S \zeta_k = -A_k \Pi_k \zeta_k$ and therefore

$$\begin{aligned}\dot{\epsilon}_k =& A_k x_k - A_k \Pi_k \zeta_k + B_k H_k (x_k - \Pi_k \zeta_k) + E_k w_k - \Pi_k \delta_k - B_k H_k (x_k - \hat{x}_{k,k}) \\ =& (A_k + B_k H_k) \epsilon_k + E_k w_k - \Pi_k \delta_k - B_k H_k e_{k,k}.\end{aligned} \tag{B.1}$$

Now, we consider the Lyapunov function

$$V(\epsilon) = \sum_{k=1}^{N} \underbrace{\epsilon_k^\top X_k \epsilon_k}_{V_k(\epsilon_k)}, \tag{B.2}$$

where X_k is the solution to (4.53). With (4.55) and (4.53), an upper bound on the Lie derivative of $V_k(\epsilon_k)$ can be obtained using the completion of squares argument:

$$\begin{aligned}\dot{V}_k =& \epsilon_k^\top \left(A_k^\top X_k + X_k A + 2 X_k B_k H_k \right) \epsilon_k + 2 \epsilon_k^\top X_k \left(E_k w_k - \Pi_k \delta_k - B_k H_k e_k^k \right) \\ \leq& \epsilon_k^\top \left(A_k^\top X_k + X_k A + 2 X_k B_k H_k \right) \epsilon_k + \epsilon_k^\top X_k \left(\frac{1}{\mu^2} E_k E_k^\top + \frac{1}{\mu^2} \Pi_k \Pi_k^\top + \frac{1}{\lambda^2} B_k B_k^\top \right) X_k \epsilon_k \\ &+ \mu^2 \left(\|w_k\|^2 + \|\delta_k\|^2 \right) + \|e_k^k\|^2_{\lambda^2 H_k^\top H_k}\end{aligned}$$

$$
\begin{aligned}
=&\epsilon_k^\top \left(A_k^\top X_k + X_k A + \frac{1}{\mu^2} X_k \left(E_k E_k^\top + \Pi_k \Pi_k^\top \right) X_k \right) \epsilon_k \\
&+ \epsilon_k^\top (\lambda H_k + \frac{1}{\lambda} B_k^\top X_k)^\top (\lambda H_k + \frac{1}{\lambda} B_k^\top X_k) \epsilon_k - \lambda^2 \epsilon_k^\top H_k^\top H_k \epsilon_k \\
&+ \mu^2 \left(\|w_k\|^2 + \|\delta_k\|^2 \right) + \|e_k^k\|^2_{\lambda^2 H_k^\top H_k}
\end{aligned}
$$

Now let H_k be defined as in (4.55). Subsequently, it holds that

$$
\begin{aligned}
\dot{V}_k \leq &\epsilon_k^\top \left(A_k^\top X_k + X_k A + \frac{1}{\mu^2} X_k \left(E_k E_k^\top + \Pi_k \Pi_k^\top \right) X_k \right) \epsilon_k - \lambda^2 \epsilon_k^\top H_k^\top H_k \epsilon_k \\
&+ \mu^2 \left(\|w_k\|^2 + \|\delta_k\|^2 \right) + \|e_k^k\|^2_{\lambda^2 H_k^\top H_k}.
\end{aligned}
$$

Using the Riccati equation (4.53) we obtain

$$
\dot{V}_k \leq -2(q_k + p_k) \epsilon_k^\top \widetilde{C}_k^\top \widetilde{C}_k \epsilon_k + \mu^2 \left(\|w_k\|^2 + \|\delta_k\|^2 \right) + \|e_k^k\|^2_{\lambda^2 H_k^\top H_k}.
$$

In the absence of perturbation, we know that $\zeta_j - \zeta_k \to 0$ and $x_k - \hat{x}_{k,k} \to 0$ exponentially fast for all $k, j \in \mathcal{N}$. Since $A_k - \frac{1}{\lambda^2} B_k B_k^\top X_k$ is Hurwitz, we can now conclude that $\epsilon_k \to 0$ for all $k = 1, ..., N$, i.e. Property 1 of Problem 5 is satisfied.

On the other hand, when the system is affected by disturbances, integrating the above inequality over the interval $[0, T]$ leads to the following inequality

$$
\begin{aligned}
&V(\epsilon(T)) + \sum_{k=1}^N \int_0^T 2(q_k + p_k) \epsilon_k^\top \widetilde{C}_k^\top \widetilde{C}_k \epsilon_k dt \\
&\leq \mu^2 \sum_{k=1}^N \int_0^T \|w_k\|^2 dt + \mu^2 \sum_{k=1}^N \int_0^T \|\delta_k\|^2 dt + \sum_{k=1}^N \int_0^T \|e_k^k\|^2_{\lambda^2 H_k^\top H_k} + \sum_{k=1}^N \epsilon_{k,0}^\top X_k \epsilon_{k,0}.
\end{aligned}
$$

Using the definition of the weight W^k (4.54) and the performance property of the distributed observer established in Property 2 of Problem 6, we can conclude

$$
\begin{aligned}
&\sum_{k=1}^N \int_0^T 2(q_k + p_k) \epsilon_k^\top \widetilde{C}_k^\top \widetilde{C}_k \epsilon_k dt \\
&\leq \mu^2 \sum_{k=1}^N \int_0^T \|w_k\|^2 dt + \mu^2 \sum_{k=1}^N \int_0^T \|\delta_k\|^2 dt \\
&\quad + \gamma^2 \sum_{k=1}^N \int_0^T (\|w^k\|^2 + \|\eta^k\|^2) dt + I_0 + \sum_{k=1}^N \epsilon_{k,0}^\top X_k \epsilon_{k,0} \\
&\leq \overline{\kappa}^2 \sum_{k=1}^N \int_0^T \|w_k\|^2 dt + \mu^2 \sum_{k=1}^N \int_0^T \|\delta_k\|^2 dt + \overline{\theta}^2 \sum_{k=1}^N \int_0^T \|\eta^k\|^2 dt + I_0 + \sum_{k=1}^N \epsilon_{k,0}^\top X_k \epsilon_{k,0},
\end{aligned}
$$

where $\overline{\kappa}^2 = \mu^2 + q_{max}\gamma^2$ and $\overline{\theta} = \gamma$.

For the left-hand-side, we note that

$$\begin{aligned}\sum_{k=1}^{N} 2(q_k + p_k)\epsilon_k^\top \widetilde{C}_k^\top \widetilde{C}_k\epsilon_k dt =& \sum_{k=1}^{N}\sum_{j\in\mathcal{N}_k} 2\left(\epsilon_k^\top \widetilde{C}_k^\top \widetilde{C}_k\epsilon_k + \epsilon_j^\top \widetilde{C}_j^\top \widetilde{C}_j\epsilon_j\right) dt \\ \geq& \sum_{k=1}^{N}\sum_{j\in\mathcal{N}_k} \left(\epsilon_k^\top \widetilde{C}_k^\top \widetilde{C}_k\epsilon_k - 2\epsilon_k^\top \widetilde{C}_k^\top \widetilde{C}_j\epsilon_j + \epsilon_j^\top \widetilde{C}_j^\top \widetilde{C}_j\epsilon_j\right) dt \\ =& \sum_{k=1}^{N}\sum_{j\in\mathcal{N}_k} \|\widetilde{C}_k\epsilon_k - \widetilde{C}_j\epsilon_j\|^2 dt\end{aligned} \tag{B.3}$$

The term $\int_0^T \|\delta_k\|^2 dt$ characterizes the effect of the transient dynamics during the time, when the controller states (4.49) are not yet synchronized. As mentioned, from (Scardovi & Sepulchre, 2009), we know that $\zeta_j - \zeta_k \to 0$ exponentially for all $k, j = 1, ..., N$, and thus, $\int_0^\infty \|\delta_k\|^2 dt$ is finite. Concluding, we have the inequality

$$\sum_{k=1}^{N}\int_0^T \|\widetilde{C}_k\epsilon_k - \widetilde{C}_j\epsilon_j\|^2 dt \leq \sum_{k=1}^{N}\int_0^T (\overline{\kappa}^2\|w_k\|^2 + \overline{\theta}^2\|\eta^k\|^2)dt + \overline{I}_0$$

with

$$\overline{I}_0 = \mu^2\sum_{k=1}^{N}\int_0^T \|\delta_k\|^2 dt + I_0 + \sum_{k=1}^{N}\epsilon_{k,0}^\top X_k\epsilon_{k,0}.$$

With $T \to \infty$ and Lemma 4.3, Property 2 of Problem 5 holds. ∎

B.2 Proof of Theorem 4.7

Proof.

The proof follows the lines of the proof of Theorem 4.6, so only the differences are pointed out. Suppose the conditions of Theorem 4.7 hold.

Again, we consider the regulation error $\epsilon_k = x_k - \Pi_k\zeta_k$. Similar to the proof of Theorem 4.6, using equations (4.70) and (4.71), the derivative of ϵ_k is found to be

$$\begin{aligned}\dot{\epsilon}_k =& \dot{x}_k - \Pi_k\dot{\zeta}_k \\ =& A_k x_k + \sum_{j\in\mathcal{N}_k} A_{kj}x_j + B_k\Gamma_k\zeta_k + B_k H_k(\hat{x}_{k,k} - \Pi_k\zeta_k) + B_k\sum_{j\in\mathcal{N}_k} H_{kj}(\hat{x}_{k,j} - \Pi_j\zeta_k) \\ &+ E_k w_k - \Pi_k S\zeta_k - \Pi_k\delta_k.\end{aligned}$$

With the regulator equations (4.67) we have

$$B_k\Gamma_k\zeta_k - \Pi_k S\zeta_k = -(A_k\Pi_k + \sum_{j\in\mathcal{N}_k} A_{kj}\Pi_j)\zeta_k \tag{B.4}$$

and therefore, the derivative of ϵ_k can be reformulated to

$$\begin{aligned}
\dot{\epsilon}_k =& A_k x_k - A_k\Pi_k\zeta_k + \sum_{j\in\mathcal{N}_k} A_{kj}x_j - \sum_{j\in\mathcal{N}_k} A_{kj}\Pi_j\zeta_k \\
&+ B_k H_k(x_k - \Pi_k\zeta_k) + B_k \sum_{j\in\mathcal{N}_k} H_{kj}(x_j - \Pi_j\zeta_k) + E_k w_k \\
&- \Pi_k\delta_k - B_k H_k(x_k - \hat{x}_{k,k}) - B_k \sum_{j\in\mathcal{N}_k} H_{kj}(x_j - \hat{x}_{k,j}) \\
=& (A_k + B_k H_k)\epsilon_k + \sum_{j\in\mathcal{N}_k} (A_{kj} + B_k H_{kj})\epsilon_j + \underbrace{\sum_{j\in\mathcal{N}_k} (A_{kj} + B_k H_{kj})\Pi_j(\zeta_j - \zeta_k)}_{\widetilde{\delta_k}} \\
&+ E_k w_k - \Pi_k\delta_k - B_k(H_k e_{k,k} + \sum_{j\in\mathcal{N}_k} H_{kj}e_{k,j}).
\end{aligned}$$

Now, we consider the Lyapunov function

$$V(\epsilon) = \sum_{k=1}^{N} \underbrace{\epsilon_k^\top X_k^{-1}\epsilon_k}_{V_k(\epsilon_k)}, \tag{B.5}$$

where X_k is the solution to (4.72), (4.73). Like in the proof of Theorem 4.6, we use the control inputs (4.71) and the completion of squares argument to find the upper bound on the Lie-derivative of $V_k(\epsilon_k)$

$$\begin{aligned}
\dot{V}_k \leq& \epsilon_k^\top \left(A_k^\top X_k^{-1} + X_k^{-1}A - \frac{1}{\lambda^2}X_k^{-1}B_k B_k^\top X_k^{-1} \right)\epsilon_k + 2\epsilon_k^\top X_k^{-1} \sum_{j\in\mathcal{N}_k} (A_{kj} + B_k H_{kj})\epsilon_j \\
&+ \epsilon_k^\top X_k^{-1}\frac{1}{\mu^2}\left(E_k E_k^\top + \Pi_k\Pi_k^\top + I_{n_k}\right)X_k^{-1}\epsilon_k + \mu^2\left(\|w_k\|^2 + \|\delta_k\|^2 + \|\widetilde{\delta}_k\|^2\right) + \|e_k\|^2_{W^k},
\end{aligned} \tag{B.6}$$

with the weighting matrix W^k defined in (4.75). The LMIs (4.72) can be transformed using the Schur-Complement into

$$Q_k + \widetilde{X}_k + X_k\left((q_k + p_k)\widetilde{C}_k^\top\widetilde{C}_k + \pi_k q_k I_{n_k}\right)X_k \leq 0.$$

With (4.74) and after multiplication with X_k^{-1} from both sides we have

$$X_k^{-1}A_k + A_k^\top X_k^{-1} - \frac{1}{\lambda^2}X_k^{-1}B_kB_k^\top X_k^{-1} + \frac{1}{\mu^2}X_k^{-1}(E_kE_k^\top + \Pi_k\Pi_k^\top + I_{n_k})X_k^{-1} \\ + X_k^{-1}\widetilde{X}_kX_k^{-1} + (q_k+p_k)\widetilde{C}_k^\top\widetilde{C}_k + \pi_kq_kI_{n_k} + \widetilde{\alpha}_kX_k^{-1} \preceq 0. \tag{B.7}$$

Moreover, multiplying the LMIs (4.73) with $\mathrm{diag}(X_k^{-1}, I_{n_{j_1}}, ..., I_{n_{j_{p_k}}})$ from both sides results in the LMIs

$$\begin{bmatrix} -X_k^{-1}\widetilde{X}_kX_k^{-1} & X_k^{-1}(A_{kj_1}+B_kH_{kj_1})-\widetilde{C}_k^\top\widetilde{C}_{j_1} & \dots & X_k^{-1}(A_{kj}+B_kH_{kj_1})-\widetilde{C}_k^\top\widetilde{C}_{j_{p_k}} \\ * & -\pi_{j_1}I_{n_{j_1}} & & \\ \vdots & & \ddots & \\ * & & & -\pi_{j_{p_k}}I_{n_{j_{p_k}}} \end{bmatrix} \preceq 0. \tag{B.8}$$

In continuation to (B.6), with (B.7) we have

$$\dot{V}_k \le -\epsilon_k^\top\left(X_k^{-1}\widetilde{X}_kX_k^{-1} + (q_k+p_k)\widetilde{C}_k^\top\widetilde{C}_k + \pi_kq_kI_{n_k}\right)\epsilon_k \\ + 2\epsilon_k^\top X_k^{-1}\sum_{j\in\mathcal{N}_k}(A_{kj}+B_kH_{kj})\epsilon_j + \mu^2\left(\|w_k\|^2 + \|\delta_k\|^2 + \|\widetilde{\delta}_k\|^2\right) + \|e_k\|_{W^k}^2,$$

and with (B.8), this results in

$$\dot{V}_k \le -\epsilon_k^\top\left((q_k+p_k)\widetilde{C}_k^\top\widetilde{C}_k + \pi_kq_kI_{n_k}\right)\epsilon_k + \sum_{j\in\mathcal{N}_k}\pi_j\epsilon_j^\top\epsilon_j + 2\epsilon_k^\top\sum_{j\in\mathcal{N}_k}\widetilde{C}_k^\top\widetilde{C}_j\epsilon_j \\ + \mu^2\left(\|w_k\|^2 + \|\delta_k\|^2 + \|\widetilde{\delta}_k\|^2\right) + \|e_k\|_{W^k}^2 - \widetilde{\alpha}_k\epsilon_k^\top X_k^{-1}\epsilon_k.$$

Summing up $\dot{V}_k, k\in\mathcal{N}$, we have the Lie-derivative

$$\dot{V} \le \sum_{k=1}^N\left(-(q_k+p_k)\epsilon_k^\top\widetilde{C}_k^\top\widetilde{C}_k\epsilon_k + 2\epsilon_k^\top\widetilde{C}_k^\top\sum_{j\in\mathcal{N}_k}\widetilde{C}_j\epsilon_j\right) \\ + \sum_{k=1}^N\mu^2\left(\|w_k\|^2 + \|\delta_k\|^2 + \|\widetilde{\delta}_k\|^2\right) + \sum_{k=1}^N\|e_k\|_{W^k}^2 - \sum_{k=1}^N\widetilde{\alpha}_k\epsilon_k^\top X_k^{-1}\epsilon_k \tag{B.9}$$

The rest of the proof then follows from the proof of Theorem 4.6. ∎

B.3 Proof of Lemma 4.5

Proof. The observer error dynamics for each agent k has been established in (4.81). Now, we use a Lyapunov function $\widetilde{V} = \sum_{k=1}^N \underbrace{e_k^\top P_k e_k}_{\widetilde{V}_k(e_k)}$, and for the components $\widetilde{V}_k(e_k)$ the Lie derivative can be calculated using (4.85) and (4.87) as

$$\begin{aligned}\dot{\widetilde{V}}_k(e_k) =& e_k^\top \left(Q^k - q_k \begin{bmatrix} \Pi_k^{-1\top} \widetilde{P}_k \Pi_k^{-1} + \Pi_k^{-1\top} \widetilde{P}_k \Pi_k^{-1} & 0 \\ 0 & 0 \end{bmatrix} - \alpha_k P_k \right) e_k + 2 e_k^\top P_k \overline{E}_k w_k \\ &+ 2 e_k^\top \left(F^k \Big[\Pi_j^{-1} e_{j,j} \Big]_{j \in \mathcal{N}_k} + \widetilde{F}^k \Big[\Delta_j e_{j,j} \Big]_{j \in \mathcal{N}_k} \right) \\ &+ 2 e_k^\top P^k \left(\begin{bmatrix} 0 \\ I_{p_k v} \end{bmatrix} - \widetilde{K}^k \right) [\Delta_j \epsilon_j]_{j \in \mathcal{N}_k} - G^k \eta_k. \end{aligned}$$

Thus, with the LMIs (4.86), we have

$$\begin{aligned}\dot{\widetilde{V}}_k(e_k) \leq& -e_k^\top \left(q_k \begin{bmatrix} \Pi_k^{-1\top} \widetilde{P}_k \Pi_k^{-1} + \Delta_k^{-1\top} \widetilde{P}_k \Delta_k^{-1} & 0 \\ 0 & 0 \end{bmatrix} + \alpha_k P_k + W^k \right) e_k \\ &+ \beta (w_k^\top \xi^k + \eta_k^\top \eta_k) + \sum_{j \in \mathcal{N}_k} e_j^{j\top} (\Pi_j^{-1\top} \widetilde{P}^j \Pi_j^{-1} + \Delta_j^\top \widetilde{P}^j \Delta_j) e_{j,j} + \frac{1}{q_{max}} \sum_{j \in \mathcal{N}_k} \epsilon_j^\top \Delta_j^\top \Delta_j \epsilon_j. \end{aligned}$$

Summing up the $\widetilde{V}_k$s, it holds for $\widetilde{V}$ that

$$\begin{aligned}\dot{\widetilde{V}} \leq& -\sum_{k=1}^N \lambda^2 e_{k,k}^\top H_k^\top H_k e_{k,k} - \alpha_k e_k^\top P^k e_k + \sum_{k=1}^N \beta (\|\xi^k\|^2 + \|\eta_k\|^2) \\ &+ \sum_{k=1}^N \frac{1}{q_{max}} \sum_{j \in \mathcal{N}_k} \epsilon_j^\top \Delta_j^\top \Delta_j \epsilon_j \\ \dot{\widetilde{V}} \leq& -\lambda^2 \sum_{k=1}^N \|H_k e_{k,k}\|^2 + \beta \sum_{k=1}^N (\|w_k\|^2 + \|\eta_k\|^2) + \beta \sum_{k=1}^N q_k \underbrace{\|\Pi_k^{-1} E_k w_k\|}_{\leq \overline{\sigma}(\Pi_k^{-1} E_k) \|w_k\|} {}^2 \\ &+ \frac{1}{q_{max}} \sum_{k=1}^N q_k \|\epsilon_k\|^2_{\Delta_k^\top \Delta_k} - \alpha_k e_k^\top P^k e_k \end{aligned} \tag{B.10}$$

Integrating both sides of (B.10) on the interval $[0, T]$, we obtain

$$\begin{aligned} -\widetilde{V}(e(0)) &+ \lambda^2 \sum_{k=1}^N \int_0^T \|H_k e_{k,k}\|^2 dt \\ &\leq \beta \sum_{k=1}^N \int_0^T \|\eta_k\|^2 dt + \beta \sum_{k=1}^N (1 + \widetilde{q}_k) \int_0^T \|w_k\|^2 dt + \sum_{k=1}^N \int_0^T \|\epsilon_k\|^2_{\Delta_k^\top \Delta_k} dt \end{aligned}$$

Letting $T \to \infty$, the proof of the Lemma is complete. ■

B.4 Proof of Theorem 4.10

Proof. Let all condition of the theorem be satisfied. We consider the regulation error $\epsilon_k = x_k - \Pi_k \xi - \widetilde{\Pi}_k \zeta_k$ and its derivative

$$\begin{aligned}\dot{\epsilon}_k =& \dot{x}_k - \Pi_k \dot{\xi} - \widetilde{\Pi}_k \dot{\zeta}_k \\ =& A_k x_k + B_k \Gamma_k \hat{\xi}_k + B_k \widetilde{\Gamma}_k \hat{\zeta}_k^k - B_k H_k \hat{\epsilon}_k + B_k^w w_k + B_k^\xi \xi + B_k^\zeta \zeta_k - \Pi_k S \xi - \widetilde{\Pi}_k \widetilde{S}_k \zeta_k \\ =& A_k x_k + B_k \Gamma_k \xi - B_k \Gamma_k (\xi - \hat{\xi}_k) + B_k \widetilde{\Gamma}_k \zeta_k - B_k \widetilde{\Gamma}_k (\zeta_k - \hat{\zeta}_k^k) \\ & - B_k H_k \epsilon_k + B_k H_k (\epsilon_k - \hat{\epsilon}_k) + B_k^w w_k + (B_k^\xi - \Pi_k S)\xi + (B_k^\zeta - \widetilde{\Pi}_k \widetilde{S}_k)\zeta_k.\end{aligned}$$

With (4.100), we have

$$\begin{aligned}\dot{\epsilon}_k =& A_k x_k - A_k \Pi_k \xi - A_k \widetilde{\Pi}_k \zeta_k - B_k H_k \epsilon_k + (B_k H_k \epsilon_k - \hat{\epsilon}_k) + B_k^w w_k - B_k \Gamma_k (\xi - \hat{\xi}_k) \\ & - B_k \widetilde{\Gamma}_k (\zeta_k - \hat{\zeta}_k^k) \\ =& (A_k - B_k H_k)\epsilon_k + B_k H_k (x_k - \hat{x}_k^k) + B_k^w w_k - B_k (\Gamma_k + H_k \Pi_k)(\xi - \hat{\xi}_k) \\ & - B_k (\widetilde{\Gamma}_k + H_k \widetilde{\Pi}_k)(\zeta_k - \hat{\zeta}_k^k)\end{aligned}$$

and with the Definition (4.104) the derivative of ϵ_k is

$$\dot{\epsilon}_k = (A_k - B_k H_k)\epsilon_k + B_k H_k (x_k - \hat{x}_k^k) + B_k^w w_k - B_k T_k (\xi - \hat{\xi}_k) - B_k \widetilde{T}_k (\zeta_k - \hat{\zeta}_k^k). \tag{B.11}$$

Now we consider the Lyapunov function

$$V(\epsilon) = \sum_{k=1}^{N} \underbrace{\epsilon_k^\top X_k \epsilon_k}_{V_k(\epsilon_k)}, \tag{B.12}$$

where X_k solves (4.102). With (B.11) the Lie-derivate is

$$\begin{aligned}\dot{V}_k =& \epsilon_k^\top \left(A_k^\top X_k + X_k A - 2 X_k B_k H_k\right) \epsilon_k + 2\epsilon_k^\top X_k B_k^d w_k \\ & + 2\epsilon_k^\top X_k B_k \left(H_k (x_k - \hat{x}_k^k) - T_k (\xi - \hat{\xi}_k) - \widetilde{T}_k (\zeta_k - \hat{\zeta}_k^k)\right)\end{aligned}$$

and by completion of squares similar to Appendix B.1, we have

$$\begin{aligned}\dot{V}_k \leq & \epsilon_k^\top \left(A_k^\top X_k + X_k A + 2 X_k B_k H_k\right) \epsilon_k + \frac{1}{\mu_k^2} \epsilon_k^\top X_k B_k^w B_k^{d\top} X_k \epsilon_k + \mu_k^2 \|w_k\|^2 \\ & \frac{3}{3\lambda_k^2} \epsilon_k^\top X_k B_k B_k^\top X_k \epsilon_k + 3\lambda_k^2 \left(\|x_k - \hat{x}_k^k\|_{H_k^\top H_k}^2 + \|\xi - \hat{\xi}_k\|_{T_k^\top T_k}^2 + \|\zeta_k - \hat{\zeta}_k^k\|_{\widetilde{T}_k^\top \widetilde{T}_k}^2\right) \\ =& \epsilon_k^\top \left(A_k^\top X_k + X_k A + \frac{1}{\mu_k^2} X_k B_k^w B_k^{d\top} X_k\right) \epsilon_k + \mu_k^2 \|w_k\|^2\end{aligned}$$

$$
\begin{aligned}
&+ \epsilon_k^\top \left(2X_k B_k H_k + \frac{1}{\lambda_k^2} X_k B_k B_k^\top X_k \right) \epsilon_k + \|e_k\|_{W_k}^2 \\
=&\epsilon_k^\top \left(A_k^\top X_k + X_k A + \frac{1}{\mu_k^2} X_k B_k^d B_k^{d\top} X_k \right) \epsilon_k \\
&+ \epsilon_k^\top (\lambda_k H_k + \frac{1}{\lambda_k} B_k^\top X_k)^\top (\lambda_k H_k + \frac{1}{\lambda_k} B_k^\top X_k) \epsilon_k - \lambda_k^2 \epsilon_k^\top H_k^\top H_k \epsilon_k \\
&+ \mu_k^2 \|w_k\|^2 + \|e_k\|_{W_k}^2 .
\end{aligned}
$$

Let H_k be defined as in (4.105). Then, we have

$$
\dot{V}_k \leq \epsilon_k^\top \left(A_k^\top X_k + X_k A + \frac{1}{\mu_k^2} X_k B_k^w B_k^{d\top} X_k \right) \epsilon_k - \lambda_k^2 \epsilon_k^\top H_k^\top H_k \epsilon_k + \mu_k^2 \|w_k\|^2 + \|e_k\|_{W_k}^2
$$

and with the Riccati Equation (4.102), this delivers the inequality

$$
\dot{V}_k \leq -\epsilon_k^\top C_k^{e\top} C_k^e \epsilon_k + \mu_k^2 \|w_k\|^2 + \|e_k\|_{W_k}^2 . \tag{B.13}
$$

Further, the regulation errors z_k are

$$
z_k = C_k^e x_k + z_k^\xi \xi + z_k^\zeta \zeta_k ,
$$

and by (4.100), this can be transformed as

$$
z_k = C_k^e x_k - C_k^e \Pi_k \xi - C_k^e \widetilde{\Pi}_k \zeta_k = C_k^e \epsilon_k . \tag{B.14}
$$

Let $w_k, \eta_k \equiv 0$ for all $k \in \mathcal{N}$, then (4.98) shows that $\lim_{t\to\infty} e_k(t) = 0$. With (B.11), (B.14) and the condition that $A_k - \frac{1}{\lambda_k^2} B_k B_k^\top X_k$ is Hurwitz, this implies that

$$
w_k, \eta_k \equiv 0 \; \forall \; k \in \mathcal{N} \Rightarrow \lim_{t\to\infty} z_k(t) = 0,
$$

which is Property 1 of Problem 7.

By integration of (B.13) over the interval $[0, T]$ we get the inequality

$$
V(T) - V(0) \leq -\sum_{k=1}^N \int_0^T \epsilon_k^\top C_k^{e\top} C_k^e \epsilon_k dt + \sum_{k=1}^N \int_0^T \max_k \mu_k^2 \|w_k\|^2 dt + \sum_{k=1}^N \int_0^T \|e_k\|_{W_k}^2 dt.
$$

And with $T \to \infty$ and (B.13),(B.14) this results in

$$
\begin{aligned}
\sum_{k=1}^N \int_0^\infty z_k^\top z_k dt \leq& \max_k \mu_k^2 \sum_{k=1}^N \int_0^\infty \|w_k\|^2 dt + \beta \sum_{k=1}^N \int_0^\infty (\|w_k\|^2 + \|\eta_k\|^2) dt \\
&+ \underbrace{J_0 + V(0) - V(t)}_{I_0} \\
=&\gamma^2 \sum_{k=1}^N \int_0^\infty \|w_k\|^2 dt + \widetilde{\gamma}^2 \sum_{k=1}^N \int_0^\infty \|\eta_k\|^2 dt + I_0 ,
\end{aligned}
$$

with $\gamma = \sqrt{\max_k \mu_k^2 + \beta}$ and $\widetilde{\gamma} = \sqrt{\beta}$, which shows Property 2 of Problem 7. ■

Bibliography

Abou-Kandil, H., Freiling, G. & Jank, G. (1993). Necessary conditions for constant solutions of coupled riccati equations in nash games. *Systems & Control Letters*, *21* (4), 295 - 306.

Abou-Kandil, H., Jank, G. & Freiling, G. (1994). Solution and asymptotic behavior of coupled riccati equations in jump linear systems. *IEEE Transactions on Automatic Control*, *39* (8), 1631–1636.

Açikmese, B. & Corless, M. (2011). Observers for systems with nonlinearities satisfying incremental quadratic constraints. *Automatica*, *47* (October), 1339–1348.

Açikmese, B., Mandic, M. & Speyer, J. L. (2014). Decentralized observers with consensus filters for distributed discrete-time linear systems. *Automatica*, *50* (4), 1037–1052.

Anta, A. & Tabuada, P. (2010). To sample or not to sample: Self-triggered control for nonlinear systems. *IEEE Transactions on Automatic Control*, *55* (9), 2030–2042.

Arcak, M. & Kokotović, P. (2001a). Feasibility conditions for circle criterion designs. *Systems & Control Letters*, *42* (5), 405 - 412.

Arcak, M. & Kokotović, P. (2001b). Nonlinear observers : a circle criterion design and robustness analysis. *Automatica*, *37*, 1923–1930.

Arcak, M., Larsen, M. & Kokotović, P. (2003). Circle and Popov criteria as tools for nonlinear feedback design. *Automatica*, *39* (4), 643 - 650.

Aström, K. J. & Kumar, P. R. (2014). Control: A perspective. *Automatica*, *50* (1), 3–43.

Baheti, R. & Gill, H. (2011). Cyber-physical systems. *The impact of control technology*, *12*, 161–166.

Barooah, P. & Hespanha, J. a. P. (2007). Estimation on graphs from relative measurements. *IEEE Control Systems Magazine* (August), 57–74.

Bernhardsson, B. & Åström, K. J. (1999). Comparison of periodic and event based sampling for first-order stochastic systems. In *14th ifac world congress*.

Bertsekas, D. & Tsitsiklis, J. N. (1989). *Parallel and Distributed Computation: Numerical Methods*. Prentice-Hall.

Bliman, P.-A. (2000). Extension of popov absolute stability criterion to non-autonomous systems with delays. *International Journal of Control, 73* (15), 1349–1361.

Blondel, V. D., Hendrickx, J. M. & Tsitsiklis, J. N. (2009). On krause's multi-agent consensus model with state-dependent connectivity. *IEEE Transactions on Automatic Control, 54* (11), 2586–2597.

Bolognani, S., Del Favero, S., Schenato, L. & Varagnolo, D. (2010). Consensusbased distributed sensor calibration and leastsquare parameter identification in wsns. *International Journal of Robust and Nonlinear Control, 20* (2), 176-193.

Borkar, V. & Varaiya, P. (1982). Asymptotic agreement in distributed estimation. *IEEE Transactions on Automatic Control, 27* (3), 650–655.

Boyd, S., Parikh, N., Chu, E., Peleato, B. & Eckstein, J. (2011). Distributed optimization and statistical learning via the alternating direction method of multipliers. *Foundations and Trends® in Machine Learning, 3* (1), 1–122.

Briat, C., Sename, O. & Lafay, J.-F. (2011). Design of LPV observers for LPV time-delay systems: an algebraic approach. *International Journal of Control, 84* (9), 1533–1542.

Callier, F. & Willems, J. (1981). Criterion for the convergence of the solution of the riccati differential equation. *IEEE Transactions on Automatic Control, 26* (6), 1232–1242.

Callier, F. & Winkin, J. (1995). Convergence of the time-invariant riccati differential equation towards its strong solution for stabilizable systems. *Mathematical Analysis and Applications* (192), 230–257.

Carli, R., Chiuso, A., Schenato, L. & Zampieri, S. (2008). Distributed Kalman filtering based on consensus strategies. *IEEE J. on Selected Areas in Comm., 26* (4), 622–633.

Carlson, D., Hershkowitz, D. & Shasha, D. (1992). Block diagonal semistability factors and Lyapunov semistability of block triangular matrices. *Linear Algebra and Its Applications*, *172* (C), 1–25.

Chen, G., Lewis, F. L. & Xie, L. (2011). Finite-time distributed consensus via binary control protocols. *Automatica*, *47* (9), 1962–1968.

Chung, W. H. & Speyer, J. L. (1995). A general framework for decentralized estimation. In *Proc. American Control Conference (ACC)* (S. 2931–2935). Seattle, Washington.

Corfmat, J.-P. & Morse, A. S. (1976). Decentralized control of linear multivariable systems. *Automatica*, *12* (5), 479–495.

D'Andrea, R. & Dullerud, G. E. (2003). Distributed control design for spatially interconnected systems. *IEEE Transactions on Automatic Control*, *48* (9), 1478–1495.

de Souza, C. E., Coutinho, D. & Kinnaert, M. (2016). Mean square state estimation for sensor networks. *Automatica*, *72*, 108 - 114.

Dieci, L. & Eirola, T. (1994). Positive definiteness in the numerical solution of riccati differential equations. *Numerische Mathematik*, *67* (3), 303–313.

Dimarogonas, D. V., Frazzoli, E. & Johansson, K. H. (2012). Distributed event-triggered control for multi-agent systems. *IEEE Transactions on Automatic Control*, *57* (5), 1291–1297.

Dimarogonas, D. V. & Johansson, K. H. (2009). Event-triggered control for multi-agent systems. In *Decision and control, 2009 held jointly with the 2009 28th chinese control conference. cdc/ccc 2009. proceedings of the 48th ieee conference on* (S. 7131–7136).

Ding, Z. (2015). Consensus disturbance rejection with disturbance observers. *IEEE Transactions on Industrial Electronics*, *62* (9), 5829-5837.

Doherty, P. & Rudol, P. (2007). A UAV search and rescue scenario with human body detection and geolocalization. In *Proc. Australasian Joint Conference on Artificial Intelligence* (S. 1–13).

Dörfler, F., Pasqualetti, F. & Bullo, F. (2013). Continuous-time distributed observers with discrete communication. *IEEE Journal of Selected Topics in Signal Processing*, *7* (2), 296–304.

Drath, R. & Horch, A. (2014). Industrie 4.0: Hit or hype? *IEEE industrial electronics magazine*, *8* (2), 56–58.

Farina, M. & Carli, R. (2016). Partition-based distributed kalman filter with plug and play features. *IEEE Transactions on Control of Network Systems*, *PP* (99), 1-1.

Fattouh, A., Sename, O. & Dion, J. (1998). $\mathcal{H}_\infty$ observer design for time-delay systems. *Proc. 37th IEEE Confer. on Decision & Control*, 4545–4546.

Fax, J. A. & Murray, R. M. (2004). Information flow and cooperative control of vehicle formations. *IEEE Transactions on Automatic Control*, *49* (9), 1465–1476.

Feingold, D. G., Varga, R. S. et al. (1962). Block diagonally dominant matrices and generalizations of the gerschgorin circle theorem. *Pacific J. Math*, *12* (4), 1241–1250.

Francis, B. A. & Wonham, W. M. (1975). The internal model principle for linear multivariable regulators. *Applied mathematics and optimization*, *2* (2), 170–194.

Francis, B. A. & Wonham, W. M. (1976). The internal model principle of control theory. *Automatica*, *12* (5), 457–465.

Frasca, P., Carli, R., Fagnani, F. & Zampieri, S. (2009). Average consensus on networks with quantized communication. *International Journal of Robust and Nonlinear Control*, *19* (16), 1787–1816.

Freiling, G., Jank, G. & Abou-Kandil, H. (1996). On global existance of solutions to coupled matrix riccati equations in closed loop nash games. *IEEE Transactions on Automatic Control*, *41* (2), 264–269.

Fridman, E. & Shaked, U. (2001). A new h filter design for linear time delay systems. *IEEE Transactions on Signal Processing*, *49* (11), 2839–2843.

Gauthier, J. P., Hammouri, H. & Othman, S. (1992). A simple observer for nonlinear systems applications to bioreactors. *IEEE Transactions on Automatic Control*, *37* (6), 1875.

Ge, X. & Han, Q.-L. (2015). Distributed event-triggered $\mathcal{H}_\infty$ filtering over sensor networks with communication delays. *Information Sciences*, *291*, 128–142.

Geiger, A., Lenz, P. & Urtasun, R. (2012). Are we ready for autonomous driving? the kitti vision benchmark suite. In *Proc. IEEE Conference on Computer Vision and Pattern Recognition (CVPR)* (S. 3354–3361).

Grip, H. F., Saberi, A. & Stoorvogel, A. A. (2015). Synchronization in networks of minimum-phase, non-introspective agents without exchange of controller states: homogeneous, heterogeneous, and nonlinear. *Automatica*, *54*, 246–255.

Grip, H. F., Yang, T., Saberi, A. & Stoorvogel, A. A. (2012a). Decentralized control for output synchronization in heterogeneous networks of non-introspective agents. In *Proc. American Control Conference (ACC)* (S. 812–819).

Grip, H. F., Yang, T., Saberi, A. & Stoorvogel, A. A. (2012b). Output synchronization for heterogeneous networks of non-introspective agents. *Automatica*, *48* (10), 2444-2453.

Haddad, W. M., Hui, Q., Chellaboina, V. & Nersesov, S. (2004). Vector dissipativity theory for discrete-time large-scale nonlinear dynamical systems. *International Journal of Control*, *77* (10), 907–919.

Isidori, A. & Byrnes, C. (1990). Output regulation of nonlinear systems. *IEEE Transactions on Automatic Control*, *35* (2), 131–140.

Isidori, A., Marconi, L. & Casadei, G. (2014). Robust output synchronization of a network of heterogeneous nonlinear agents via nonlinear regulation theory. *IEEE Transactions on Automatic Control*, *59* (10), 2680–2691.

Jadbabaie, A. & Morse, A. (2003). Coordination of groups of mobile autonomous agents using nearest neighbor rules. *IEEE Transactions on Automatic Control*, *48* (9), 1675–1675.

Ji, M. & Egerstedt, M. B. (2007). Distributed coordination control of multi-agent systems while preserving connectedness. *IEEE Transactions on Robotics*, *23* (4), 693–703.

Jodar, L. & Hervas, A. (1989). An algorithm for solving coupled differential matrix systems: approximations, convergence and upper error bounds. *Journal of Computational and Applied Mathematics*, *25* (3), 351 - 362.

Jonsson, U. & Megretski, A. (2000). The zames-falb iqc for systems with integrators. *IEEE Transactions on Automatic Control, 45* (3), 560–565.

Kalman, R. E. (1960). A new approach to linear filtering and prediction problems. *Transactions of the ASME–Journal of Basic Engineering, 82* (Series D), 35–45.

Kalman, R. E. (1962). Canonical structure of linear dynamical systems. *Proceedings of the National Academy of Sciences, 48* (4), 596–600.

Kalman, R. E. (1963). Lyapunov functions for the problem of lur'e in automatic control. *Proceedings of the national academy of sciences, 49* (2), 201–205.

Kalman, R. E. & Bucy, R. S. (1961). New results in linear filtering and prediction theory. *Journal of Basic Engineering, 83* (1), 95–108.

Kamal, A. T., Farrell, J. A. & Roy-Chowdhury, A. K. (2013). Information weighted consensus filters and their application in distributed camera networks. *IEEE Transactions on Automatic Control, 58* (12), 3112–3125.

Khalil, H. K. (2001). *Nonlinear systems (3rd edition)*. Prentice Hall.

Khalil, H. K. & Praly, L. (2014). High-gain observers in nonlinear feedback control. *International Journal of Robust and Nonlinear Control, 24*, 993–1015.

Kim, H., Shim, H. & Seo, J. H. (2011). Output consensus of heterogeneous uncertain linear multi-agent systems. *IEEE Transactions on Automatic Control, 56* (1), 200–206.

Kim, J., Shim, H. & Wu, J. (2016). On distributed optimal kalman-bucy filtering by averaging dynamics of heterogeneous agents. In *Proc. 55th IEEE Conference on Decision and Control (CDC)*.

Knobloch, H. W., Isidori, A. & Flockerzi, D. (1993). *Topics in control theory, chapter 1 (the problem of output regulation)*. Birkhäuser.

Kuka AG. (2016). *Industrie 4.0 - glossary.*

Langbort, C., Chandra, R. S. & D'Andrea, R. (2004). Distributed control design for systems interconnected over an arbitrary graph. *IEEE Transactions on Automatic Control, 49* (9), 1502–1519.

Langbort, C. & Ugrinovskii, V. (2010). Diagonal stability of stochastic systems subject to nonlinear disturbances and diagonal $\mathcal{H}_2$ norms. In *Proc. 49th IEEE Conference on Decision and Control (CDC)* (S. 3188–3193).

Listmann, K. D., Wahrburg, A., Strubel, J., Adamy, J. & Konigorski, U. (2011). Partial-state synchronization of linear heterogeneous multi-agent systems. In *Proc. 50th IEEE Conference on Decision and Control (CDC)* (S. 3440–3445). Orlando, FA.

Löfberg, J. (2004). YALMIP : a toolbox for modeling and optimization in MATLAB. In *Proc. IEEE Int. Symposium on Computer Aided Control Systems Design* (S. 284 – 289).

Luenberger, D. (1966). Observers for multivariable systems. *IEEE Transactions on Automatic Control, 11* (2), 190-197.

Lunze, J. (2011). An internal-model principle for the synchronisation of autonomous agents with individual dynamics. In *Proc. 50th IEEE Conference on Decision and Control (CDC)* (S. 2106-2111). Orlando, FA.

Manyika, J., Chui, M., Bisson, P., Woetzel, J., Dobbs, R., Bughin, J. & Aharon, D. (2015). Unlocking the potential of the internet of things. *McKinsey Global Institute.*

Markoff, J. (2010). Google cars drive themselves, in traffic. *The New York Times, 10* (A1), 9.

Materassi, D. & Salapaka, M. V. (2011). A generalized zames-falb multiplier. *IEEE Transactions on Automatic Control, 56* (6), 1432–1436.

Mathew, N., Smith, S. L. & Waslander, S. L. (2015). Planning paths for package delivery in heterogeneous multirobot teams. *IEEE Transactions on Automation Science and Engineering, 12* (4), 1298–1308.

Mazo, M. & Tabuada, P. (2011). Decentralized event-triggered control over wireless sensor/actuator networks. *IEEE Transactions on Automatic Control, 56* (10), 2456–2461.

Mesbahi, M. & Egerstedt, M. (2010). *Graph Theoretic Methods in Multiagent Networks.* Princeton University Press.

Moreau, L. (2005). Stability of multiagent systems with time-dependent communication links. *IEEE Transactions on Automatic Control, 50* (2), 169–182.

Narendra, K. S. (2014). *Frequency domain criteria for absolute stability*. Elsevier.

Necoara, I., Nedelcu, V. & Dumitrache, I. (2011). Parallel and distributed optimization methods for estimation and control in networks. *Journal of Process Control, 21* (5), 756–766.

Olfati-Saber, R. (2005). Distributed kalman filter with embedded consensus filters. In *Proc. 44th IEEE Conference on Decision and Control (CDC)* (S. 8179–8184).

Olfati-Saber, R. (2006). Distributed kalman filtering and sensor fusion in sensor networks. *Networked Embedded Sensing and Control*, 157–167.

Olfati-Saber, R. (2007). Distributed Kalman filtering for sensor networks. In *Proc. 46th IEEE Conference on Decision and Control (CDC)* (S. 5492–5498). New Orleans, LA, USA: Ieee.

Olfati-Saber, R., Fax, J. A. & Murray, R. M. (2007). Consensus and cooperation in networked multi-agent systems. *Proceedings of the IEEE, 95* (1), 215–233.

Olfati-Saber, R. & Murray, R. (2004). Consensus problems in networks of agents with switching topology and time-delays. *IEEE Transactions on Automatic Control, 49* (9), 1520–1533.

Oshman, Y. & Bar-Itzhack, I. (1985). Eigenfactor solutions of the matrix riccati equation-a continuous square root algorithm. *IEEE Transactions on Automatic Control, 30* (10), 971–978.

Pittet, C., Tarbouriech, S. & Burgat, C. (1997, Dec). Stability regions for linear systems with saturating controls via circle and popov criteria. In *Proc. 36th IEEE Conference on Decision and Control (CDC)* (Bd. 5, S. 4518-4523 vol.5).

Popov, V. M. (1961). On absolute stability of non-linear automatic control systems. *Avtomatika i telemekhanika, 22* (8), 857–875.

Rabbat, M. & Nowak, R. (2004). Distributed optimization in sensor networks. In *Proc. 3rd Int. Symposium on Information Processing in Sensor Networks* (S. 20–27).

Raffard, R. L., Tomlin, C. J. & Boyd, S. P. (2004). Distributed optimization for cooperative agents: application to formation flight. In *Proc. 43rd IEEE Conference on Decision and Control (CDC)* (Bd. 3, S. 2453–2459).

Raghavan, S. & Hedrick, J. K. (1994). Observer design for a class of nonlinear systems. *International Journal of Control*, *59* (2), 515–528.

Rantzer, A. (1996). On the kalmanyakubovichpopov lemma. *Systems & Control Letters*, *28* (1), 7–10.

Ren, W., Beard, R. & Atkins, E. (2007). Information consensus in multivehicle cooperative control. *Control Systems Magazine,*, 71–82.

Ren, W., Beard, R. W. & Atkins, E. M. (2005). A survey of consensus problems in multi-agent coordination. In *Proc. american control conference (acc)* (S. 1859–1864).

Reynolds, C. W. (1987). Flocks, herds and schools: A distributed behavioral model. *ACM SIGGRAPH computer graphics*, *21* (4), 25–34.

Scardovi, L. & Sepulchre, R. (2009). Synchronization in networks of identical linear systems. *Automatica*, *45* (11), 2557–2562.

Schenato, L. (2008). Optimal estimation in networked control systems subject to random delay and packet drop. *IEEE Transactions on Automatic Control*, *53* (5), 1311–1317.

Schenato, L., Sinopoli, B., Franceschetti, M., Poolla, K. & Sastry, S. S. (2007). Foundations of Control and Estimation Over Lossy Networks Foundations of Control and Estimation Over Lossy Networks. *Proceedings of the IEEE*, *95* (1), 163–187.

Scherer, C. W. (2002). Structured finite-dimensional controller design by convex optimization. *Linear Algebra and its Applications*, *351-352*, 639–669.

Seuret, A., Dimarogonas, D. V. & Johansson, K. H. (2008). Consensus under communication delays. In *2008 47th ieee conference on decision and control* (S. 4922-4927).

Seyboth, G. S. & Allgöwer, F. (2014). Synchronized model matching: a novel approach to cooperative control of non-linear multi-agent systems. In *Proc. 19th IFAC World Congress* (S. 1985-1990).

Seyboth, G. S., Dimarogonas, D. V. & Johansson, K. H. (2013). Event-based broadcasting for multi-agent average consensus. *Automatica*, *49* (1), 245–252.

Seyboth, G. S., Ren, W. & Allgöwer, F. (2016). Cooperative control of linear multi-agent systems via distributed output regulation and transient synchronization. *Automatica*, *68*, 132–139.

Seyboth, G. S., Schmidt, G. S. & Allgöwer, F. (2012). Output synchronization of linear parameter-varying systems via dynamic couplings. In *Proc. 51st IEEE Conference on Decision and Control (CDC)* (S. 5128–5133).

Shi, D., Chen, T. & Shi, L. (2014). Event-triggered maximum likelihood state estimation. *Automatica, 500* (1), 247 - 254.

Šiljak, D. D. (1978). *Large-scale dynamic systems: stability and structure* (Bd. 2). North Holland.

Šiljak, D. D. (1991). *Decentralized Control of Complex Systems*. Dover.

Šiljak, D. D., Stipanovic, D. M. & Zecevic, A. I. (2002). Robust decentralized turbine/governor control using linear matrix inequalities. *IEEE Transactions on Power Systems, 17* (3), 715–722.

Šiljak, D. D. & Vukcevic, M. (1976). Decentralization, stabilization, and estimation of large-scale linear systems. *IEEE Transactions on Automatic Control, 21* (3), 363–366.

Skogestad, S. & Postlethwaite, I. (2007). *Multivariable feedback control: analysis and design* (Bd. 2). Wiley New York.

Stanković, S. S., Stipanović, D. M. & Šiljak, D. D. (2007). Decentralized dynamic output feedback for robust stabilization of a class of nonlinear interconnected systems. *Automatica, 43* (5), 861–867.

Su, Y. & Huang, J. (2012a). Cooperative output regulation of linear multi-agent systems. *IEEE Transactions on Automatic Control, 57* (4), 1062-1066.

Su, Y. & Huang, J. (2012b). Cooperative output regulation of linear multi-agent systems by output feedback. *Systems & Control Letters, 61* (12), 1248 - 1253.

Subbotin, M. V. & Smith, R. S. (2009). Design of distributed decentralized estimators for formations with fixed and stochastic communication topologies. *Automatica, 45* (11), 2491 - 2501.

Sundaram, S. & Hadjicostis, C. N. (2007). Finite-time distributed consensus in graphs with time-invariant topologies. In *Proc. American Control Conference (ACC)* (S. 711–716).

Suykens, J., Vandewalle, J. & De Moor, B. (1998). An absolute stability criterion for the lur'e problem with sector and slope restricted nonlinearities. *IEEE Transactions on circuits and systems, Part 1: Fundamental Theory and Applications*, *45*, 1007–1009.

Swaroop, D. & Hedrick, J. (1999). Constant spacing strategies for platooning in automated highway systems. *Journal of Dynamic Systems, Measurement, and Control*, *121* (3), 462–470.

Swigart, J. (2010). *Optimal Controller Synthesis for Decentralized Systems* (Unveröffentlichte Dissertation). Stanford University.

Tabuada, P. (2007). Event-triggered real-time scheduling of stabilizing control tasks. *IEEE Transactions on Automatic Control*, *52* (9), 1680–1685.

Thau, F. E. (1973). Observing the state of non-linear dynamic systems. *International Journal of Control*, *17* (March 2013), 471–479.

Trentelman, H. L., Takaba, K. & Monshizadeh, N. (2013). Robust synchronization of uncertain linear multi-agent systems. *IEEE Transactions on Automatic Control*, *58* (6), 1511–1523.

Tuna, E. (2008). LQR-based coupling gain for synchronization of linear systems. *http://arxiv.org/pdf/0801.3390.pdf*.

Ugrinovskii, V. (2011). Distributed robust filtering with H_∞ consensus of estimates. *Automatica*, *47* (1), 1–13.

Ugrinovskii, V. (2013). Distributed robust estimation over randomly switching networks using consensus. *Automatica*, *49* (1), 160–168.

Ugrinovskii, V. & Langbort, C. (2011). Distributed $\mathcal{H}_\infty$ consensus-based estimation of uncertain systems via dissipativity theory. *IET Control Theory & Applications*, *5* (12), 1458–1469.

Veenman, J. & Scherer, C. W. (2011). Iqc-synthesis with general dynamic multipliers. *IFAC Proceedings Volumes*, *44* (1), 4600–4605.

Vicsek, T., Czirók, A., Ben-Jacob, E., Cohen, I. & Shochet, O. (1995). Novel type of phase transition in a system of self-driven particles. *Physical review letters*, *75* (6), 1226.

Wahrburg, A. & Adamy, J. (2011). Observer-based synchronization of heterogeneous multi-agent systems by homogenization. In *Proc. Australian Control Conference (AuCC)* (S. 386–391).

Wang, L. & Morse, A. S. (2017). A distributed observer for a time-invariant linear system. *IEEE Transactions on Automatic Control, PP* (99).

Wang, L. & Xiao, F. (2010). Finite-time consensus problems for networks of dynamic agents. *IEEE Transactions on Automatic Control, 55* (4), 950–955.

Wang, X., Hong, Y. & Jiang, Z.-p. (2010). A distributed control approach to a robust output regulation problem for multi-agent linear systems. *IEEE Trans. Automat. Control, 55* (12), 2891–2895.

Wang, X. & Lemmon, M. D. (2009). Self-triggered feedback control systems with finite-gain stability. *IEEE Transactions on Automatic Control, 54* (3), 452–467.

Wieland, P. (2010). *From Static to Dynamic Couplings in Consensus and Synchronization among Identical and Non-Identical Systems*. PhD thesis. University of Stuttgart.

Wieland, P. & Allgöwer, F. (2009). An internal model principle for synchronization. In *Proc. IEEE Int. Conference on Control and Automation* (S. 285–290). Christchurch, NZ.

Wieland, P., Kim, J.-S. & Allgöwer, F. (2011). On topology and dynamics of consensus among linear high-order agents. *International Journal of Systems Science, 42* (10), 1831–1842.

Wieland, P., Sepulchre, R. & Allgöwer, F. (2011). An internal model principle is necessary and sufficient for linear output synchronization. *Automatica, 47* (5), 1068–1074.

Wieland, P., Wu, J. & Allgöwer, F. (2013). On synchronous steady states and internal Models of diffusively coupled systems. *IEEE Transactions on Automatic Control, 58* (10), 2591–2602.

Willems, J. (1973). The circle criterion and quadratic Lyapunov functions for stability analysis. *IEEE Transactions on Automatic Control, 18* (2), 184-184.

Wu, J. & Allgöwer, F. (2012). A constructive approach to synchronization using relative information. In *Proc. 51st IEEE Conference on Decision and Control (CDC)* (S. 5960 – 5965). Maui, HI.

Wu, J. & Allgöwer, F. (2016a). Distributed nonlinear observer with robust performance - A circle criterion approach. *arXiv:1604.03014*.

Wu, J. & Allgöwer, F. (2016b). Verteilte Zustandsschätzung zur Ausgangsregulierung von verteilten Systemen mit gekoppelten Messgrößen. *at-Automatisierungstechnik*, *64* (8), 645-657.

Wu, J., Elser, A., Zeng, S. & Allgöwer, F. (2016). Consensus-based distributed kalman-bucy filter for continuous-time systems. *IFAC-PapersOnLine*, *49* (22), 321–326.

Wu, J., Li, L., Ugrinovskii, V. & Allgöwer, F. (2015). Distributed filter design for cooperative h-infinity-type estimation. In *Proc. IEEE Conference on Control Applications (CCA)* (S. 1373–1378).

Wu, J., Ugrinovskii, V. & Allgöwer, F. (2014). Cooperative estimation for synchronization of heterogeneous multi-agent systems using relative information. In *Proc. 19th IFAC World Congress* (S. 4662–4667).

Wu, J., Ugrinovskii, V. & Allgöwer, F. (2015). Cooperative $\mathcal{H}_\infty$-estimation for large-scale interconnected linear systems. In *Proc. American Control Conference (ACC)* (S. 2119–2124).

Wu, J., Ugrinovskii, V. & Allgöwer, F. (2016). Observer-based synchronization with relative measurements and unknown neighbour models. In *Proc. Australian Control Conference (AuCC)*. Newcastle, Australia.

Wu, J., Ugrinovskii, V. & Allgöwer, F. (2017). Cooperative estimation and robust synchronization of heterogeneous multi-agent systems with coupled measurements. *IEEE Trans. Control of Networked Systems*. (In print)

Xiao, L., Boyd, S. & Kim, S.-J. (2007). Distributed average consensus with least-mean-square deviation. *Journal of Parallel and Distributed Computing*, *67* (1), 33–46.

Xu, D., Hong, Y. & Wang, X. (2014). Distributed output regulation of nonlinear multi-agent systems via host internal model. *IEEE Transactions on Automatic Control*, *59* (10), 2784–2789.

Yu, L. & Wang, J. (2013). Robust cooperative control for multi-agent systems via distributed output regulation. *Systems & Control Letters*, *62* (11), 1049–1056.

Zamani, M. & Ugrinovskii, V. (2014). Minimum-energy distributed filtering. In *Proc. 53rd IEEE Conference on Decision and Control (CDC)* (S. 3370–3375).

Zames, G. & Falb, P. (1968). Stability conditions for systems with monotone and slope-restricted nonlinearities. *SIAM Journal on Control*, *6* (1), 89–108.

Zeitz, M. (1987). The extended luenberger observer for nonlinear systems. *Systems & Control Letters*, *9* (2), 149 - 156.

Zelazo, D. & Mesbahi, M. (2008). On the observability properties of homogeneous and heterogeneous networked dynamic systems. In *Proc. 47th IEEE Conference on Decision and Control (CDC)* (S. 2997–3002).

Zelazo, D. & Mesbahi, M. (2011). Graph-theoretic analysis and synthesis of relative sensing networks. *IEEE Transactions on Automatic Control*, *56* (5), 971–982.

Zhang, W., Branicky, M. S. & Phillips, S. M. (2001). Stability of networked control systems. *IEEE Control Systems*, *21* (1), 84–99.

Zhu, M. & Martínez, S. (2010). Discrete-time dynamic average consensus. *Automatica*, *46* (2), 322–329.